Weight Wat
New Complete Cookbook
#2022-23

MW01686994

1000 Day Easy & Delicious Recipes for Weight Loss | Smart Point™ Edition, Loss Weight and Improve Your Lifestyle

Alice Bennett

Weight Watchers New Complete Cookbook #2022-23

Table of Contents

Introduction

US News recently conducted a study to determine which of several well-known diets is the most popular and effective. Unsurprisingly, the Weight Watchers program was selected as one of the most popular methods for losing weight and controlling one's calorie intake. The Weight Watchers program teaches you to track and limit your food portions to avoid overeating. In this instance, the premise is apparent! Each week, you receive a certain number of SmartPoints based on your gender, body weight, age, and other factors. This week, you may consume as much food as you wish without exceeding your daily limit. SmartPoints are assigned to the various food groups and ingredients that contribute to a balanced, healthy lifestyle. Since it's founding 50 years ago, Weight Watchers has undergone considerable change. The Freestyle version of the diet is likely the best because it combines the best aspects of all the other versions into a single, straightforward plan. The first chapter discusses the Weight Watchers program, and the remainder of the book contains delicious and comforting recipes. I hope you enjoy the book and can strictly adhere to your Weight Watchers Freestyle diet. Enjoy your day, and keep smiling!

What is Weight Watchers?

Weight Watchers is a 1960s-established diet program. It presently has 4,7 million users worldwide, including 350,000 in the United Kingdom.

Currently, Weight Watchers has been rebranded as 'WW' with the motto 'Wellbeing that Works,' shifting the emphasis away from weight reduction and toward a healthy approach to weight loss, wellness, and body positivity.

The new moniker is part of a massive rebranding of the diet brand. It also involves significant alterations to its philosophy, which covers mental health and total wellbeing for the first time and weight reduction.

Other recent WW ambassadors include Kate Hudson, Robbie Williams, Curtis Pritchard, Alison Hammond, and Vicky Pattison.

What Is the Core of Weight Watchers?

Unlike other diet plans that restrict your caloric intake and dictate what you can and cannot eat, this diet plan can help you lose weight safely and effectively. Weight Watchers can function independently. It recognizes that you have a lot going on in your life and won't be able to purchase expensive or hard-to-find ingredients for remaining safe. This diet emphasizes Smart Points, which will urge you to consume the healthiest meals, but it also rewards good items and discourages unhealthy ones. This diet emphasizes Smart Points, which will guide you to finish the most nutritious meals, but it also rewards good things and discourages unhealthy ones. This approach emphasizes awareness of one's eating habits.

Each day, you will be granted a set number of points to utilize, and you will have the option of how to use them. Different point values will be assigned to the meals you select, and you may even create your recipes and calculate the point values. This program encourages you to cheat during the week if you truly want to or can't resist a large celebration. You'll find that you may count items toward your daily point totals while consuming them. As long as you make informed decisions for the remainder of the day, these little infractions will not detract from your hard work. Other parts of Weight Watchers include considering the healthy foods you may consume over the week. Examples include attending meetings and adding more physical exercise to daily life.

Weight Watchers meetings are a distinguishing characteristic that has led to the organization's expansion. These sessions are designed to hold you accountable for the predicted weight loss. These are often hosted weekly and enable you to meet with others in your region pursuing the same objectives. You will enter and be discreetly weighed, and you will then be able to meet with others to learn new things and find the motivation to continue working. The quantity of physical activity you participate in while on the Weight Watchers program is crucial.

Although this has improved slightly since the program's beginning, engaging in various physical activities is still essential to maintain a healthy body and shed as much weight as feasible. Amazingly, Weight Watchers may be used to transform one's lifestyle radically.

This is not a program that requires arduous effort on challenging activities to attain the intended outcomes. Weight Watchers encourages you to change your lifestyle, such as eating the right foods, getting adequate exercise, and learning the proper information, to stay on track and attain the intended outcomes.

How does Weight Watchers Work?

Weight Watchers is a great diet plan that may assist you in losing weight in a healthy and controllable manner. There are, however, diets that are more concerned with informing you precisely what you may and cannot consume. While this technique may work for some, others may find it difficult to give up their favorite foods. Furthermore, it may be challenging to shop for food when adhering to specific diet regimens.

The Weight Watchers system will be unique. It realizes that you don't have much free time to go out and buy pricey items or look for hard-to-find components to keep your health. The Smart Points method used in this article allows you to eat the most convenient meals and encourages and rewards the consumption of healthy foods while discouraging the consumption of unhealthy ones. Paying attention to how you eat is the key to success on this diet. Each day, you will be given a specific number of points to utilize how you see fit. You may create recipes and calculate the point values for the meals you prepare. If you genuinely want to cheat or can't resist a big party, you can stick to this schedule and still have some fun during the week.

They will contribute to your daily point totals without losing flavor or delight. If you make solid decisions for the remainder of the day, these tiny transgressions will not detract from your efforts.

Weight Watchers and other diet programs need more than just preparing nutritious meals for the week. Meetings and making additional physical modifications to your routine are two examples. Weight Watchers' hallmark meeting structure is one of the critical reasons for the organization's success. These sessions are designed to hold you responsible for the weight reduction goals you've set for yourself. These weekly events provide you to meet folks who share your interests and improve your network. Come in, get your weight taken privately, and then attend a group session to discuss what you've learned and gain encouragement from others. The degree of physical activity you maintain is one of the most critical parts of the Weight Watchers program. Even though this has altered slightly since the program's inception, it is still vital to go out there and engage in a range of activities to maintain your body healthy and assist you in losing weight. Weight Watchers is fantastic because it may be utilized to create long-term adjustments to one's lifestyle. With this program, you won't have to spend a lot of time and effort on things that are difficult to finish. Weight Watchers suggests changing your lifestyle in various ways, including eating healthier, increasing your physical activity, and educating yourself on how to succeed.

Zero Point Food List

Zero Point foods are ones that you can eat without tracking or measuring. They don't count against your daily Personal Point total and are an excellent choice when you're hungry but don't want to jeopardize your progress.

Weight Watchers used to give a comprehensive general list of Zero Point meals. Still, now that the new Personal Points system has been implemented, everyone's list will be distinct, depending on your specific profile and dietary needs.
Joining the WW program and answering the short set of questions created by their team of nutrition experts is the best way to find out what you'll get to eat.

This enables them to understand what kind of foods you love eating the most, allowing them to assist you in creating a meal list that is specific to you.

Here's a brief list of some meals that may be on your Zero Points eating list:

- Avocados
- Eggs
- Tofu
- Corn
- Potatoes
- Tempeh
- Fruits
- Noodles/Pasta (Whole-wheat)
- Poultry
- Non-starchy vegetables
- Cottage cheese (low-fat/fat-free)
- Yogurt (low-fat/fat-free)
- Beans, lentils, peas
- Popcorn
- Fish/Shellfish
- Brown Rice

- Whole Grains
- Starchy vegetables

Breakfast Recipes

Breakfast Wraps

Prep Time: 5 Minutes

Cook Time: 5 Minutes

Servings: 4

Ingredients:

- 3/4 cup Jimmy Dean's turkey sausage crumbles or diced fully cooked ham
- 6 eggs or 1 1/2 cups liquid egg substitute
- 4 Laughing Cow Cheese Wedge, any flavor
- 4 low point tortilla
- 2 Tbsp. unsweetened almond or skim milk
- Favorite veggies

Instructions:

1. Whisk together the eggs, milk, and a pinch of salt and pepper in a small bowl.
2. Cook and stir over medium heat until the eggs are completely set, spraying a pan with nonstick cooking spray.
3. Each tortilla should have one cheese slice down the center. Cooked turkey crumbles should be added.
4. Spread the spread cheese and turkey crumbs with the egg mixture; roll up the tortilla by tucking in the edges and rolling from the bottom up.
5. Serve right away. Roll up any leftover breakfast rolls in plastic wrap or foil and place in a resalable plastic bag to freeze.

Banana Split

Prep Time: 5 Minutes

Cook Time: 0 Minutes

Servings: 1

Ingredients:

- 1-½ teaspoons sliced almonds
- ¼ cup blueberries
- 1 banana, sliced in half lengthwise
- 1 tablespoon granola
- ⅓ cup low-fat cottage cheese
- ¼ cup raspberries
- ½ teaspoon honey

Instructions:

1. Place the banana slices on a dish or in a bowl.
2. Serve with cottage cheese, raspberries, and blueberries on top.
3. Top with cereal and almonds.
4. Drizzle with honey and serve immediately.

Berry Breakfast Tacos

Prep Time: 5 Minutes

Cook Time: 2 Minutes

Servings: 2

Ingredients:

- 1½ tbsp Low sugar chocolate spread approx 30g - depending on brand
- 1 cup blueberries 100g
- 1 banana
- Calorie controlled spray oil
- 2 tortilla wraps see info on best low point wraps
- 1 cup raspberries 125g

Instructions:

1. Heat a big frying pan or skillet over low to medium heat, spritzed with calorie-controlled spray oil.
2. Heat the first tortilla wrap in the pan for about 30 seconds on each side. You want the wrap to be warmed through and slightly crisped. Recycle the leftover tortilla wrap.
3. Distribute the chocolate spread evenly between the two wrappers. The heat from the tortilla wrap should cause it to spread nicely and become a bit runny. If it doesn't, you may always warm it up in the microwave for 5 seconds before using it.
4. Serve with sliced bananas, blueberries, and raspberries on top.
5. Enjoy!

Overnight Betty Oats

Prep Time: 10 Minutes

Cook Time: 0 Minutes

Servings: 1

Ingredients:

- 1-ounce dry oats rolled
- 1 teaspoon Acai powder
- 1 ounce fresh blueberries
- 2/3 cups skim milk
- ¼ cup fresh raspberries
- ¼ teaspoon vanilla bean extract
- ½ banana, medium-sized

Instructions:

1. In a jug, mix together the acai powder, milk, and vanilla. Mash the raspberries and banana with a fork.
2. Fill an airtight container halfway with bananas, raspberries, blueberries, milk mix, and oats.
3. Make certain that it is completely covered.
4. Place it in the freezer overnight.
5. Serve and have fun!

Tasty Eggplant Casserole

Prep Time: 5 Minutes

Cook Time: 40 Minutes

Servings: 4

Ingredients:

- ¾ cup non-fat cottage cheese
- 1 cup spaghetti sauce
- 2 tablespoons parmesan cheese, grated
- ¾ cup low-fat mozzarella cheese, shredded
- ¾ pound eggplant slice, ½ inch thick

Instructions:

1. Prepare an 8x8 baking dish with nonstick cooking spray.
2. 5 to 10 minutes on the grill should be sufficient time for the eggplant to become tender
3. Place the eggplant on the pan's bottom.
4. Spread spaghetti sauce evenly over eggplant.
5. Combine mozzarella and cottage cheese in a bowl.
6. Spread it on top of sauce
7. Sprinkle with parmesan cheese and bake at 350 degrees for 30 minutes.
8. Serve right away and enjoy!

Berry-Licious Almond Smoothie

Prep Time: 10 Minutes

Cook Time: 0 Minutes

Servings: 1

Ingredients:

- 1 piece of banana
- Water as needed
- 1 cup blueberries of frozen
- 1 tablespoon of almond butter
- ½ cup almond milk

Instructions:

1. Add the items specified to the blender.
2. Blend them till they have a smooth consistency.
3. Add water to the smoothie to thin it out.
4. Serve cold and relish!

Granola Packed Bowl

Prep Time: 5 Minutes

Cook Time: 25 Minutes

Servings: 6

Ingredients:

- 5 ounces strawberries, blueberries, and raspberries
- 4 pots Caramel Layered Fromage Frais
- ¼ ounce Desiccated coconut
- 1-ounce Porridge oats

- ¼ ounce pumpkin seeds
- Cooking spray as needed
- 4 medium Bananas
- ¼ ounce sunflower seeds
- 2 teaspoons Maple Syrup
- ¼ ounce Chia seeds, dry

Instructions:

1. Preheat the oven to 30o degrees Fahrenheit
2. Prepare a baking sheet by lining it with baking paper.
3. Combine oats, maple syrup, and seeds in a large bowl.
4. Spread the mixture on the baking sheet
5. Spray with coconut oil
6. 20 minutes while stirring constantly
7. Sprinkle coconut after 15 minutes
8. Remove them from the oven and place the mixture on a second, cool baking sheet.
9. Take a bowl and arrange sliced bananas and Fromage Frais in it.
10. Sprinkle granola on top
11. Serve and delight!

French Toast Sticks

Prep Time: 10 Minutes

Cook Time: 6 Minutes

Servings: 24 Sticks

Ingredients:

- 8 slices sandwich bread
- 3 large eggs
- 2/3 cup skim milk
- 1 teaspoon vanilla extract
- ½ teaspoon powdered sugar
- A pinch of salt
- 1 teaspoon ground cinnamon

Instructions:

1. I used kitchen shears to cut each piece of bread into three equal strips and laid them aside.
2. Whisk the eggs, milk, vanilla essence, cinnamon, and salt until incorporated, then pour into a shallow dish (I use a pie pan).
3. Medium-heat a pan or griddle that has been lightly sprayed with cooking spray. Coat the bread slices with the egg mixture after dipping them into it. Allow any extra liquid to drain from the strips before placing them on the pan/griddle. Cook for two to three minutes per side, or until the sticks are golden and faintly crisp. Transfer to a dish and dust with confectioners' sugar.

Strawberry Smoothie

Prep Time: 5 Minutes

Cook Time: 0 Minutes

Servings: 2

Ingredients:

- 2 cups frozen strawberries
- 1 ½ tablespoons maple syrup, honey, or agave syrup
- 1 banana
- 1 cup milk
- ¼ cup Greek yogurt
- ½ cup ice

Instructions:

1. Place all of the ingredients in a blender, breaking up the banana. Blend until smooth and foamy, scraping the sides as needed. Garnish with a frozen strawberry and mint leaf, if preferred. Serve immediately or refrigerate in a sealed container for up to 2 days.

Spinach Smoothie

Prep Time: 5 Minutes

Cook Time: 0 Minutes

Servings: 2

Ingredients:

- 1 large green apple
- 1 tablespoon pure maple syrup

- 1 ½ tablespoons fresh squeezed lemon juice (1/2 lemon)
- ¼ cup raw cashews (or 1 banana and 1 ½ cups frozen mango or pineapple, skip the maple)
- ½ cup water
- 4 cups baby spinach leaves or chopped spinach, loosely packed
- 10 ice cubes

Instructions:

1. Leave the apple peel on while you dice the apple into bits. Blend all ingredients in a blender until smooth, pausing and scraping as needed. (If you're utilizing the fruit option, consider mixing frozen fruit in two batches.)

Mango Smoothie

Prep Time: 5 Minutes

Cook Time: 0 Minutes

Servings: 2

Ingredients:

- 3 cups frozen mango
- 1/2 banana
- ½ cup water
- 1 tablespoon fresh squeezed lemon juice
- ½ cup milk
- 5 ice cubes

Instructions:

1. Add all ingredients, liquids first, to a blender, except for 2 cups of frozen mango. Blend till creamy.

2. Add the 2 cups of frozen mango and mix again until smooth, if required adding a splash of more water or milk.

Kale Smoothie

Prep Time: 5 Minutes

Cook Time: 0 Minutes

Servings: 2

Ingredients:

- 1 large green apple
- 1 tablespoon fresh squeezed lemon juice
- 10 ice cubes
- ½ cup water
- 2 cups baby kale leaves, loosely packed
- 1 banana
- 2 cups frozen pineapple or mango

Instructions:

1. Chopped apple with the peel intact. Place the banana in the blender along with the strawberries. Add water and young kale leaves to the pan. Blend till creamy.
2. Add ice, frozen mango or pineapple, and lemon juice. Blend one more until smooth.

Bacon Quiche with Cheese

Prep Time: 10 Minutes

Cook Time: 30 Minutes

Servings: 4

Ingredients:

- 3 and a ½ ounce Broccoli, raw and cut up into small florets
- 3 and ½ ounce soft cheese, low-fat
- 2 medium spring onions, trimmed and sliced
- ½ cup skimmed milk
- 3 rasher bacon medallions, raw and roughly chopped up
- 2 and ½ ounce cheese, low fat and grated
- 4 sprays Calorie controlled cooking spray
- 2 raw whole eggs, medium sized

Instructions:

1. Preheat oven to 356 degrees Fahrenheit
2. Add broccoli to a pan and heat.
3. Cook broccoli in water that is boiling over moderate heat.
4. Prepare for 4 minutes
5. Layers should be drained and let to dry on kitchen paper.
6. Place a nonstick skillet over moderate heat.
7. Using frying spray, grease the pan and add bacon and spring onions
8. four to five minutes
9. Spray a 7-inch round, shallow cake with cooking spray.
10. Spread bacon and broccoli mixture over the foundation and sprinkle with cheese
11. Season with black pepper then whisk together the eggs, milk, and soft cheese.

12. Bake the batter in the pan for 25 minutes.
13. Serve and delight!

Toasted Guac

Prep Time: 10 Minutes

Cook Time: 5 Minutes

Servings: 4

Ingredients:

- ½ medium avocado, finely chopped
- 1 cup beans, cooked
- 1 tablespoon lime juice
- 4 slices multigrain/ sourdough, toasted
- 2 tablespoons coarsely chopped fresh coriander
- 2 medium tomato, finely fresh

Instructions:

1. In a small bowl, combine beans, tomato, coriander, and lime juice. Season with salt and pepper.
2. Spread the bean mixture on bread.
3. Serve and delight!

Simple Almond Cereal

Prep Time: 10 Minutes

Cook Time: 0 Minutes

Servings: 3

Ingredients:

- 2 tablespoons almond, chopped
- Handful of berries
- 1/3 cup water
- 1 small banana, chopped
- 2 tablespoons pepitas, roasted
- 1 tablespoon chia seeds
- 1/3 cup coconut milk

Instructions:

1. Add chia seeds and coconut milk to a bowl and set it aside for 5 minutes.
2. Take your food processor and pulse half of the pepitas and almonds.
3. Combine this with chia seeds.
4. Add water then mix
5. Add the remaining pepitas, bananas, and blueberries to the dish.
6. Combine and serve
7. Enjoy!

Perfect Bake Eggs

Prep Time: 10 Minutes

Cook Time: 20 Minutes

Servings: 4

Ingredients:

- ¼ cup fresh basil, chopped
- 2 cups marinara sauce
- 4 pieces large eggs
- ½ a cup parmesan, grated
- ¼ teaspoon red pepper flakes

Instructions:

1. Preheat the oven to 350 degrees Fahrenheit
2. Put the oven rack in the center position.
3. Take a baking dish and pour in marinara sauce
4. Sprinkle some basil over it
5. Create four thin wells and break one egg into each well.
6. The eggs are seasoned with parmesan cheese and red pepper flakes.
7. around 18 to 20 minutes
8. Serve and delight!

Tri-Ingredient Pancake

Prep Time: 15 Minutes

Cook Time: 10 Minutes

Servings: 1

Ingredients:

- 1 ripe banana, small sized
- 2 tablespoon of whole meal self-rising flour
- Cooking spray
- 1 egg, medium sized

Instructions:

1. In a medium-sized mixing basin, combine bananas.
2. Using a fork, mash them.
3. Whisk in the egg and flour, then thoroughly blend everything.
4. Allow the mixture to stand for 5 minutes.
5. Grease a nonstick frying pan with olive oil.
6. Heat the pan over medium heat and add the spoon. 2 teaspoons batter
7. Cook for two minutes on each side.
8. Repeat for the remainder of the batter.
9. Serve and have fun!

Soup and Stews

Cauliflower Leek Soup

Prep Time: 15 Minutes

Cook Time: 1 Hours

Servings: 12

Ingredients:

- 3 tablespoons butter
- 3 cloves garlic, finely chopped
- 2 tablespoons olive oil
- 3 leeks, cut into 1 inch pieces
- 1 large head cauliflower, chopped
- salt and freshly ground black pepper to taste
- 8 cups vegetable broth
- 1 cup heavy cream

Instructions:

1. Over medium heat, melt the butter and olive oil in a large pot. Cook and stir for about 10 minutes while adding leeks, cauliflower, and garlic. Stir in the vegetable broth and bring to a boil. Lower the heat, cover, and let it simmer for 45 minutes.
2. Take the soup off the heat and use an immersion blender to blend it until it is smooth. Salt and pepper it, then add heavy cream and keep blending until it's smooth.

Delightful Turmeric Soup

Prep Time: 5 Minutes

Cook Time: 25 Minutes

Servings: 4

Ingredients:

- 2 and ½ cups carrots, chopped
- ½ teaspoon cinnamon
- 1 teaspoon lemon juice
- 1 teaspoon garlic powder
- 3.5-ounce walnuts
- 1 teaspoon turmeric
- 5 teaspoons vegan stock
- 1 teaspoon chipotle paste
- ½ inch fresh ginger
- ½ teaspoon cumin

Instructions:

1. Put the ingredients on the list into a stockpot.
2. Put it on a medium-high flame.
3. Simmer for 25 minutes on low heat.
4. Use an immersion blender to mix the whole thing.
5. Serve, and have fun!

Cauliflower-Cheese Soup

Prep Time: 20 Minutes

Cook Time: 25 Minutes

Servings: 4

Ingredients:

- ¾ cup water
- 3 cups milk
- 4 ounces shredded Cheddar cheese
- ¼ cup chopped onion
- 1 cup cauliflower, chopped
- 1 cup cubed potatoes
- ½ cup finely chopped celery
- ½ cup diced carrots
- ¼ cup butter
- ¼ cup all-purpose flour
- salt and pepper to taste

Instructions:

1. Mix water, cauliflower, potatoes, celery, carrots, and onion together in a large pot. Boil for 5 to 10 minutes, or until tender. Set aside.
2. In a separate pot over medium heat, melt the butter. Cook for 2 minutes after adding the flour. Take it off the heat and slowly stir in the milk. Bring back to heat and cook until it gets thick. Mix in the vegetables and cooking liquid, then add salt and pepper to taste. Stir in the cheese until it melts, then take it off the heat.

Spicy Black Bean and Quinoa Soup

Prep Time: 15 Minutes

Cook Time: 37 Minutes

Servings: 4

Ingredients:

- 1 teaspoon olive oil
- 1 carrot, chopped
- 2 (15 ounce) cans low-sodium chicken broth
- 1 cup frozen corn
- ½ cup water
- 1 clove garlic, minced
- ½ cup chopped onion
- 1 (15 ounce) can diced tomatoes with juice
- 1 (15 ounce) can black beans, drained
- ½ cup quinoa
- 1 teaspoon ground paprika
- ½ teaspoon red pepper flakes
- 2 stalks celery, chopped
- salt and ground black pepper to taste

Instructions:

1. In a large saucepan, heat the olive oil over medium-high heat. Add the onion, celery, and carrot. Cook and stir for 3 to 5 minutes, until the onion is clear and the celery and carrot are crisp-tender. Stir in the garlic, paprika, and red pepper flakes, and cook for about a minute, or until the food smells good.

2. In the saucepan, add chicken broth, water, and quinoa. Bring to a simmer, then turn the heat down to medium-low, cover, and cook for about 20 minutes, or until the quinoa is soft.
3. Mix in the black beans, corn, and diced tomatoes with juice. Bring back to a simmer and cook, uncovered, for about 5 minutes, or until the black beans are hot all the way through. Add salt and pepper to taste.

Chicken Taco Soup

Prep Time: 5 Minutes

Cook Time: 30 Minutes

Servings: 10

Ingredients:

- 2 cups chicken, cubed and shredded
- 1 poblano pepper, diced
- 2 tablespoons taco seasoning
- 6 cups chicken broth, fat-free
- 1 cup tomato sauce
- 1 onion, diced
- 2 tomatoes, chopped
- 1 tablespoon garlic, minced
- 1 and ½ cups kidney beans
- 1 tablespoon olive oil
- 1 bell pepper, diced

Instructions:

1. Add onion, poblano pepper, bell pepper, and tomato into a large-sized stockpot

2. They took 5 minutes to cook over medium-high heat.
3. Mix in the chicken, tomato sauce, beans, broth, and spices.
4. Turn the heat down to medium
5. Let it simmer for 30 minutes
6. Add sour cream and cheese to the top.
7. Serve, and have fun!

Kimchee Soup

Prep Time: 15 Minutes

Cook Time: 25 Minutes

Servings: 4

Ingredients:

- 2 cups chopped kimchi
- 1 cup water
- 1 (12 ounce) can fully cooked luncheon meat
- 1 (3 ounce) package enoki mushrooms
- ½ pound pork belly, cut into bite-size pieces
- ¼ cup kochujang
- 1 (12 ounce) package silken tofu
- 1 tablespoon brown sugar
- 4 cloves garlic, minced

Instructions:

1. Over medium heat, combine kimchi, water, sugar, and garlic in a large pan. Bring to a simmer. Add pork belly, luncheon meat, and kochujang to the pan. Add tofu and mushrooms, making sure not to break up the tofu. Reduce heat to low-medium. 20 to 30 minutes, or until the kimchi is softened and the pork is slightly pink in the center.

Simple Potato Soup

Prep Time: 5 Minutes

Cook Time: 30 Minutes

Servings: 6

Ingredients:

- 1 pound sweet potato, peeled and cut up into medium pieces
- Freshly chopped parsley
- 2 cloves garlic, crushed
- ¼ teaspoon chili, grounded
- 2 cups chicken stock
- Low fat crème fraiche
- ¼ teaspoon cinnamon, grounded
- 1 large-sized onion, chopped
- ½ teaspoon coriander, grounded
- 2 tablespoons extra virgin olive oil
- ¼ teaspoon salt
- Coriander as needed
- ½ teaspoon ground cumin

Instructions:

2. Olive oil is added to a big pan.
3. Place it over moderate heat.
4. Sauté the onions until they are gently browned.
5. Reduce the heat to medium and then stir in the garlic.
6. Cook for a further 2 to 3 minutes, then add the sweet potato
7. Sauté for 3-4 minutes
8. Add the remaining seasonings and a pinch of salt.
9. Prepare for 2 minutes

10. Pour stock and increase the heat
11. Bring the mixture to a boil
12. Stir gently
13. Lower the lid and cover the pot, then bring the liquid to a moderate simmer.
14. Cook the potatoes for 20 minutes, or until they are soft.
15. Take the pan off the heat.
16. Utilizing an immersion blender, purée the entire mixture.
17. If the soup is too thick, add a little water.
18. Give a crème fraiche swirl.
19. Garnish with chopped parsley
20. Serve and delight!

Greek Lemon Chicken Soup

Prep Time: 25 Minutes

Cook Time: 40 Minutes

Servings: 16

Ingredients:

- 8 cups chicken broth
- 6 tablespoons chicken soup base
- ¼ cup all-purpose flour
- ½ cup fresh lemon juice
- ½ cup chopped celery
- ¼ teaspoon ground white pepper
- ¼ cup margarine
- ½ cup shredded carrots
- ½ cup chopped onion

- 1 cup cooked white rice
- 1 cup diced, cooked chicken meat
- 8 egg yolks
- 16 slices lemon

Instructions:

1. In a large saucepan, combine chicken broth, lemon juice, carrots, onions, celery, soup base, and white pepper. Bring to a boil over high heat before reducing heat and simmering for 20 minutes.
2. In a separate dish, combine margarine and flour; gradually add into soup mixture. Stir constantly while simmering for 10 minutes.
3. In the meantime, beat egg yolks in a basin until they are pale in color. Whisk in some hot soup gradually. Add the egg mixture to the saucepan and bring to a boil.
4. Add rice and chicken; heat until well heated. To serve, ladle soup into bowls and top with lemon wedges.

Saffron Coconut Broth

Prep Time: 20 Minutes

Cook Time: 25 Minutes

Servings: 4

Ingredients:

- 1 brown onion, medium sized and finely chopped
- 1 tablespoon sunflower oil
- 1/8 teaspoon saffron, dried
- 1 fresh garlic, crushed
- 1 teaspoon fresh ginger, finely grated
- 1 thinly sliced whole fresh red chili

- ½ a teaspoon coriander, grounded
- 1 medium sized fresh lemon wedges

Instructions:

1. Place a big saucepan over a heat source of medium intensity.
2. Add oil and let it to warm up.
3. Add onion and continue stirring for five minutes
4. Add coriander, chile, ginger, garlic, and saffron to the dish.
5. Prepare for 1 minute
6. Combine coconut milk, 1 cup of water, and fish sauce.
7. Bring it to a boil, then add the pumpkin and potato.
8. Reduce the heat and bring the liquid to a simmer.
9. Simmer for 8-10 minutes
10. Add beans and peppers, then cover
11. Simmer for 5 minutes more
12. Add bell pepper, bean, and cover
13. Prepare food for 5 minutes
14. Add half of the fresh coriander.
15. Lemon wedge to garnish
16. Serve and delight!

Egg Drop Soup

Prep Time: 5 Minutes

Cook Time: 5 Minutes

Servings: 5

Ingredients:

- ¼ cup green onion, thinly sliced
- 1 egg, lightly beaten

- ½ teaspoon soy sauce
- 4 cups chicken broth, low sodium
- ½ cup frozen, green pea
- ½ cup chicken breast, cooked, boneless, chopped and skinless

Instructions:

1. Place a saucepan over medium heat.
2. Soy sauce and chicken stock
3. Bring the mixture to a boil, then add the chicken, peas, and green onions.
4. Give it a good shake.
5. Re-bring the mixture to a boil.
6. Turn off the heat and softly trickle in the egg.
7. Allow a minute for the egg to set before gently stirring it.
8. Serve hot and enjoy!

Fancy Chicken Noodle Soup

Prep Time: 15 Minutes

Cook Time: 15 Minutes

Servings: 4

Ingredients:

- 1 cup thin spaghetti, broken up into 2 inch long pieces
- 8 ounces' chicken breast, cut up into bite sized portions
- Black pepper as needed
- 8 pieces' green onion, thinly sliced
- ½ cup carrots, chopped up
- ¼ teaspoon thyme, dried

- 5 cups chicken broth, fat free
- ½ cup celery, chopped
- ¼ teaspoon parsley, dried
- Salt as needed

Instructions:

1. Season your chicken in a bowl with salt and pepper.
2. Preheat a nonstick skillet over medium-high heat.
3. Cooking spray should be used.
4. Cook the chicken for 3 to 5 minutes, stirring regularly.
5. Fill a soup saucepan halfway with cooked chicken, carrots, spaghetti, chicken broth, green onion, celery, thyme, and parsley.
6. Bring the mixture to a boil, then reduce the heat.
7. Allow it to cook until your pasta is soft.
8. Season with salt and black pepper to taste.
9. Serve and have fun!

Lovely Onion Soup

Prep Time: 5 Minutes

Cook Time: 25 Minutes

Servings: 4

Ingredients:

- 2 onions, large sized and finely sliced
- 4 slices stale bread
- 4 ounce grated strong cheese
- 1 teaspoon brown sugar
- 2 tablespoons olive oil

- 1 ounce grated parmesan
- 1 cup red wine
- 2 cups vegetable stock
- 1 teaspoon herbs de Provence
- 1 tablespoon plain flour
- 1-ounce butter
- 1 measure brandy
- Salt Pepper as needed

Instructions:

1. Place a pan over medium-high heat, add oil and butter, and let to cook.
2. Cook until the sugar melts and the onions are somewhat caramelized, then add the onion and sugar.
3. Pour in the brandy and mix thoroughly. Add the flour and herbs de Provence.
4. Give it a good swirl.
5. Add the stock first, then the red wine gradually.
6. Season with salt and pepper and simmer for 20 minutes on low heat.
7. If necessary, add a little more water to thin it out.
8. Place slices of stale bread over top and sprinkle with cheese.
9. Garnish with parmesan and melted cheese if desired.
10. Serve and have fun!

Avgolemono Chicken Soup

Prep Time: 15 Minutes

Cook Time: 1 Hours 10 Minutes

Servings: 6

Ingredients:

- 1 (3 pound) whole chicken
- 3 large eggs, beaten
- 2 medium lemons, juiced
- ½ cup uncooked white rice
- salt and freshly ground black pepper to taste

Instructions:

1. Rinse the chicken and remove any internal organs. Place in a saucepan large enough to allow for movement but not so large that the broth becomes watery. Add the chicken neck if it came with the bird.
2. Fill the pot halfway with water, approximately 1 inch deep. Bring to a boil, covered. When it begins to boil, turn the heat to low and continue to cook for 45 minutes to an hour, scraping the fat from the top as it accumulates. The flesh should readily separate from the bones.
3. Place the chicken in a large mixing dish and leave aside to chill.
4. Season the rice in the broth with salt and pepper. Simmer over low heat for 20 minutes, or until rice is soft.
5. While the rice is cooking, mix together the eggs and lemon juice in a bowl. Separate the chicken flesh from the bones.

6. Turn off the heat when the rice is done. To prevent the eggs from curdling, carefully whisk in one ladle of boiling broth. Whisk in additional broth gradually until the egg mixture is warm. Return the heated egg mixture to the pot, whisking vigorously. The end product should be a creamy, cloudy soup. As required, season with salt, pepper, or lemon juice.
7. Serve the soup with salted chicken on the side or add chunks of chicken to the broth before serving.

Cabbage Soup

Prep Time: 12 Minutes

Cook Time: 30 Minutes

Servings: 8

Ingredients:

- 1⁄2yellow onion
- 1⁄2 cup chopped carrot
- 1 tablespoon tomato paste3 cups nonfat beef broth (reduced sodium)
- 2 garlic cloves, minced
- 1⁄2 teaspoon basil
- 2 cups chopped cabbage
- 1⁄2 cup green beans
- 1⁄2 cup chopped zucchini yellow or green
- 1⁄2 teaspoon oregano
- salt & pepper

Instructions:

1. Coat the pan with nonstick cooking spray. Cook for 5 minutes with the onions, carrots, and garlic.

2. To taste, add broth, tomato paste, cabbage, green beans, basil, oregano, and salt and pepper.
3. Simmer for 5-10 minutes, or until all veggies are cooked, then add the zucchini and continue to cook for another 5 minutes.
4. I experimented with several versions. Green beans are not included. In addition to the yellow onion, add chopped green onions.
5. Everything is excellent. You may use whichever veggies you choose.
6. For instructions on how to use your Slow Cooker or Instant Pot, see the page above.
7. 12 cup serving size

WW Taco Soup

Prep Time: 10 Minutes

Cook Time: 20 Minutes

Servings: 8

Ingredients:

- 16 ounces Extra-Lean 99% Ground Turkey
- 1 ounce pkg dry ranch seasoning
- 15 ounce can petite diced tomatoes
- 15 ounce can black beans rinsed and drained
- 1 medium yellow onion diced
- 15 ounce can kidney or red beans rinsed and drained
- 15 ounce can corn drained
- 10 ounce can Rotel tomatoes and green chilies
- 1 ounce pkg dry taco seasoning
- 14.5 ounce can chicken broth

Instructions:

1. Brown the ground turkey and chopped onion in a large stockpot. Remove any extra fat. Stir in the spice mixtures before adding the other ingredients. Stir everything together well, then cover and boil for 20 minutes, stirring periodically.
2. Serve immediately. Keep any leftovers in an airtight jar in the refrigerator. Individual portions may also be frozen.

Tortellini Soup

Prep Time: 15 Minutes

Cook Time: 15 Minutes

Servings: 4

Ingredients:

- 8-10 oz. package Tortellini pasta-dried, frozen, or fresh
- 1 Tbsp. olive oil
- 1 tsp. Basil
- 1 tsp. Oregano
- 1 tsp. Black pepper
- ¼ cup Parmesan cheese
- 1 ½ cups fresh spinach,chopped
- 2 stalks celery, diced
- 1 small onion, diced
- 3 cloves garlic,crushed
- 1 tsp. Rosemary
- 3 carrots, diced
- 1-14 oz. can diced tomatoes
- 1 cup water

- 3 cups chicken broth

Instructions:

1. In a large saucepan over medium-high heat, sauté the olive oil, onions, carrots, and celery until the onions turn translucent. Stir in the garlic until well cooked.
2. Stir in the chopped tomatoes, broth, water, basil, oregano, rosemary, and black pepper; cook for 5 minutes, stirring periodically.
3. Cook for a further 15 minutes, stirring often, after adding the tortellini pasta, spinach, and Parmesan cheese.
4. OR, in your crock pot, combine onions, carrots, celery, garlic, broth, water, oregano, rosemary, and black pepper. Cook for 3-4 hours on low heat. Cook for another hour on low with the tortellini, spinach, and Parmesan cheese.
5. Serve and enjoy this 3 point soup per cup!

Weight Watchers Instant Pot Beef Stew

Prep Time: 15 Minutes

Cook Time: 40 Minutes

Servings: 8

Ingredients:

- 1 lb lean beef cubes
- 3 cups beef broth
- ½ teaspoon of dried oregano
- 1 bay leaf
- 1 cup celery chopped
- 1 medium onion chopped

- 1 cups red potatoes cubed with skins on or off
- Salt and pepper to taste
- 2 cloves garlic minced
- 1 15 ounce can tomato sauce
- 1 cup carrots chopped
- 1 teaspoon Olive oil
- 1 cup corn frozen, fresh or canned--drained

Instructions:

1. On the Instant Pot, choose Sauté and add the olive oil, garlic, onions, dried oregano, and meat, browning it for approximately 4 minutes.
2. Sauté the carrots, celery, salt, and pepper for another 4 to 5 minutes.
3. Turn off the Sauté function and add the remaining ingredients to the Instant Pot.
4. Press the Manual button for 17 minutes and then let it go naturally. This procedure took around 25-30 minutes in my pot. I want my meat to be really soft. If required, increase the cooking time to ensure the meat is tender.

Weight Watchers Slow Cooker Irish Stew Recipe

Prep Time: 30 Minutes

Cook Time: 8 Hours

Servings: 8

Ingredients:

- 1 cup beef broth
- 3 carrots, peeled and cut into chunks

- 3 tablespoons all-purpose flour
- 2 cups frozen peas
- 2 pounds' boneless leg of lamb, trimmed and cut into 1-inch cubes
- 8 small red potatoes, scrubbed and cut in half
- 1 cup chopped onion
- 14 ounces canned diced tomatoes
- 1 cup Irish stout (such as Guinness)
- 2 parsnips, peeled and quartered lengthwise and cut into 1-inch slices
- ¼ teaspoon black pepper
- 1 small turnip, peeled, halved and each half cut into 8 chunks
- 1 bay leaf
- ½ teaspoon salt
- ¼ cup chopped flat leaf parsley

Instructions:

1. 5-quart slow cooker is the ideal size.
2. In the crock pot, combine the broth, stout, lamb, potatoes, tomatoes, onions, parsnips, carrots, turnips, bay leaf, salt, and pepper.
3. Cook on LOW for 6 to 8 hours, or until the lamb is fork-tender and the veggies are soft.
4. To thicken the stew, combine the flour and 12 cup water in a mixing bowl (or shake in a closed container) until smooth. Add to the stew.
5. Cook, covered, on HIGH for 20 minutes, or until thickened.
6. Before serving, remove the bay leaf and toss in the peas and parsley.

WW Beef Stew in The Slow Cooker

Prep Time: 15 Minutes

Cook Time: 45 Minutes

Servings: 1.5 Cup

Ingredients:

- 1 1/2 pounds' lean beef cut into 1" cubes and trimmed of fat
- 2 cups carrots, peeled and cut into 2" pieces
- 2 tablespoons diced garlic
- 2 tablespoons flour
- 3 tablespoons olive oil
- 1 tablespoon low sodium Worcestershire sauce
- 1 teaspoon Kitchen Bouquet
- 2 cups parsnips, peeled and cut into 2" pieces
- 1 cup of crushed tomatoes
- 2 cups low sodium beef broth
- 1 small bag of frozen peas or mixed vegetables
- 1 medium diced onion
- 1 tablespoon low sodium soy sauce
- Salt and pepper

Instructions:

1. In a small mixing dish, combine the meat and 1 tablespoon oil.
2. Toss in the flour, salt, and pepper to coat the meat.
3. In a large skillet, heat the remaining olive oil. Cook for 2 minutes, or until the garlic is aromatic (this step is not shown in the images since we did not use onions).

4. Brown the meat for a further 2 minutes, but do not let it to cook all the way through.
5. Steps 8-16: Combine the ingredients listed above in your slow cooker.
6. Combine the beef broth, soy and Worcestershire sauces, Kitchen Bouquet, onion powder, tomatoes, carrots, and parsnips in a mixing bowl.
7. Combine thoroughly.
8. For 4.5 hours, set the slow cooker on high. When the timer goes off, make sure the steak is tender. Cook for another 30 minutes on high with the frozen veggies.

Vegetable and Salad

WW Roasted Vegetables

Prep Time: 10 Minutes

Cook Time: 30 Minutes

Servings: 8

Ingredients:

- summer squash, halved lengthwise
- 12 ounces canned artichoke hearts, without oil, drained
- 1 teaspoon dried thyme
- olive oil flavored cooking spray, divided (4 sprays)
- zucchini, halved lengthwise
- 1 medium green pepper, seeded and quartered
- 1⁄4 teaspoon black pepper, freshly ground
- 1 cup red onion, sliced
- 1⁄2 teaspoon table salt
- 2 medium sweet red peppers, seeded and quartered

Instructions:

1. Preheat your oven to 450oF. 2 big baking sheets, sprayed with cooking spray
2. Place peppers, squash, zucchini, onion, and artichokes on prepared baking sheets, flesh side up, and spray with cooking spray; season with thyme, salt, and black pepper.
3. Cook for 20 to 25 minutes, or until the veggies are soft and golden brown. Serve the veggies in bite-size pieces. Each serving yields around 3/4 cup.

Vegetable Beef Soup

Prep Time: 10 Minutes

Cook Time: 5 Hours 30 Minutes

Servings: 10

Ingredients:

- ¼ teaspoon garlic powder, or to taste
- 1 (15.25 ounce) can whole kernel corn, drained
- 1 (14.5 ounce) can tomato sauce
- 2 (14 ounce) cans beef broth
- 1 ½ pounds beef stew meat, cut into 1/2-inch cubes
- ¼ tablespoon onion powder, or to taste
- 1 (15 ounce) can green beans, drained
- 1 (46 fluid ounce) bottle tomato-vegetable juice cocktail
- 1 (6 ounce) can tomato paste
- salt and ground black pepper, to taste

Instructions:

1. In a large saucepan over medium heat, combine beef broth and beef stew meat. Bring broth to a boil, then lower to a low heat and cook meat until tender, approximately 45 minutes.
2. Combine the meat, corn, green beans, tomato sauce, and tomato paste in a mixing bowl. Season the tomato-vegetable juice cocktail with garlic powder, onion powder, salt, and pepper. Bring the mixture to a boil, then lower to a low heat for 5 hours.

Basic Vegetable Stock

Prep Time: 15 Minutes

Cook Time: 35 Minutes

Servings: 12

Ingredients:

- 2 large carrots
- 1 large onion
- 8 sprigs fresh parsley
- 2 large bay leaves
- 1 tablespoon olive oil
- 6 sprigs fresh thyme
- 2 stalks celery, including some leaves
- 8 cloves garlic, minced
- 1 bunch green onions, chopped
- 1 teaspoon salt
- 2 quarts water

Instructions:

1. 1 inch slices of carrot, celery, and onion
2. In a soup pot, heat the oil over high heat. Carrots, celery, onion, scallions, garlic, parsley, thyme, and bay leaves should be added now. Cook, stirring regularly, for 5 to 10 minutes, or until veggies soften and begin to brown.
3. Bring the salt and water to a boil. Reduce the heat to low and cover for 30 minutes.

4. Strain. Remove the veggies.

Summer Vegetable Ratatouille

Prep Time: - Minutes

Cook Time: - Minutes

Servings: 8

Ingredients:

- 1 chopped red bell pepper
- 4 roma (plum) tomatoes, chopped
- ½ cup olive oil
- 1 bay leaf
- 4 sprigs fresh thyme
- 2 medium yellow squash, cubed
- 2 green bell peppers, seeded and cubed
- 1 yellow bell pepper, diced
- 2 onion, sliced into thin rings
- 3 cloves garlic, minced
- 2 tablespoons chopped fresh parsley
- 1 medium eggplant, cubed
- 2 zucchini, cubed
- salt and pepper to taste

Instructions:

1. In a large saucepan over medium-low heat, heat 1 1/2 tbsp of the oil. Cook until the onions and garlic are tender.

2. Heat 1 1/2 tablespoon olive oil in a large pan and sauté the zucchini in batches until nicely browned on both sides. Place the zucchini in the saucepan with the onions and garlic.
3. Saute all of the remaining veggies one batch at a time, adding 1 1/2 tbsp olive oil to the pan each time. Once each set of vegetables has been sautéed, add them to the main pot as in step 2.
4. Season with salt and pepper to taste. Cover the saucepan with the bay leaf and thyme. Cook for 15 to 20 minutes on medium heat.
5. Cook for another 10-15 minutes after adding the chopped tomatoes and parsley to the main saucepan. Stir seldom.
6. Remove the bay leaf and season to taste.

Tuna and Vegetable Casserole

Prep Time: 20 Minutes

Cook Time: 30 Minutes

Servings: 8

Ingredients:

- 1 (10.75 ounce) can condensed cream of mushroom soup
- 1 tablespoon Worcestershire sauce
- 1 cup bite-size summer squash pieces
- 1 cup frozen peas
- ½ (12 ounce) package egg noodles
- 1 cup sliced carrots
- 1 cup 1-inch pieces fresh-cut green beans
- 1 cup sliced fresh mushrooms
- 1 (7 ounce) can albacore tuna, drained and broken into chunks
- ½ onion, chopped

- ½ cup shredded Cheddar cheese
- ⅓ cup milk
- 1 teaspoon olive oil
- 1 cup French-fried onions
- salt and ground black pepper to taste

Instructions:

1. Warm up a big saucepan of gently salted water. Cook egg noodles in boiling water for 5 minutes, tossing regularly, until cooked through yet firm to the biting. Drain.
2. Preheat the oven to 425°F (220 degrees C).
3. In a microwave-safe glass or ceramic dish, combine green beans, squash, peas, and carrots; heat until soft, 3 to 10 minutes. Drain and rinse veggies with cool water.
4. In a pan over medium-high heat, heat the olive oil; sauté the mushrooms in the heated oil until soft, 5 to 7 minutes.
5. In a large mixing bowl, combine mushroom soup and milk; stir in veggies, mushrooms, tuna, onion, and Cheddar cheese. Add Worcestershire sauce, salt, and black pepper to taste. Fill a casserole dish halfway with the mixture and top with the French fried onions.
6. Bake until bubbling, approximately 20 minutes in a preheated oven.

Cobb Salad

Prep Time: 20 Minutes

Cook Time: 10 Minutes

Servings: 6

Ingredients:

- 1 (8 ounce) bottle Ranch-style salad dressing
- 3 cups chopped, cooked chicken meat
- 2 tomatoes, seeded and chopped
- 1 head iceberg lettuce, shredded
- 3 eggs
- 3 green onions, chopped
- 6 slices bacon
- 1 avocado - peeled, pitted and diced
- ¾ cup blue cheese, crumbled

Instructions:

1. Cover eggs fully with cold water in a saucepan; come to a boil, then cover and remove from heat. Allow eggs to rest for 10 to 12 minutes before peeling and chopping.
2. Cook the bacon in a big, deep pan while the eggs are cooking. Cook for 7 to 10 minutes over medium-high heat, or until uniformly brown. Set aside after draining and crumbling.
3. Divide the shredded lettuce among the plates. Stack bacon, eggs, chicken, tomatoes, blue cheese, green onions, and avocado on top. Dressing to taste.

Best Chicken Salad

Prep Time: 5 Minutes

Cook Time: 5 Minutes

Servings: 2

Ingredients:

- 2 cups chopped, cooked chicken meat
- ½ cup mayonnaise
- ¼ teaspoon ground black pepper
- ½ cup blanched slivered almonds
- 1 stalk celery, chopped
- 1 tablespoon lemon juice

Instructions:

1. In a frying pan, toast the nuts. Toast, shaking often, over medium-high heat. They burn readily, so keep an eye on them.
2. In a medium mixing bowl, combine mayonnaise, lemon juice, and pepper. Toss in the chicken, almonds, and celery.

Seafood Salad Supreme

Prep Time: 30 Minutes

Cook Time: 30 Minutes

Servings: 8

Ingredients:

- 1-pound fresh shrimp, peeled and deveined
- 2 tablespoons chopped fresh parsley

- 1 cup shredded mozzarella cheese
- 1 tablespoon butter
- 1 cup mayonnaise
- 1-pound crabmeat
- 1 red bell pepper, chopped
- 1 green onion, thinly sliced
- 1 (8 ounce) package seashell pasta
- ½ teaspoon Old Bay Seasoning TM
- 3 tablespoons fresh lemon juice
- 2 strips celery, sliced
- ¼ cup slivered, toasted almonds

Instructions:

1. Melt the butter in a large pan over medium heat and sauté the shrimp until pink. Cook for another minute, or until the crabmeat is cooked through. Place aside.
2. Warm up a big saucepan of gently salted water. Cook for 8 to 10 minutes, or until pasta is al dente; drain.
3. Allow the shrimp and crabmeat to cool to room temperature in a large mixing dish. Incorporate the spaghetti, celery, bell pepper, and green onion. Mix in the mozzarella cheese, toasted almonds, mayonnaise, and lemon juice. Season with parsley and old bay seasoning, to taste. To properly combine, toss everything together.
4. Refrigerate for at least two hours before serving.

Crab Pasta Salad

Prep Time: 10 Minutes

Cook Time: 10 Minutes

Servings: 12

Ingredients:

- 2 tablespoons chopped fresh chives
- 1 (8 ounce) package imitation crabmeat, coarsely chopped - or more to taste
- 1 (10 ounce) package frozen peas, thawed
- 1 (12 ounce) package rotini pasta
- 1 (8 ounce) can water chestnuts, chopped
- 1 cup reduced-fat mayonnaise
- 1 tablespoon chopped fresh dill

Instructions:

1. Bring a large saucepan of lightly salted water to a boil; cook the rotini for 8 minutes, or until cooked but firm to the bite; drain and rinse under cool water.
2. In a large mixing bowl, combine peas, water chestnuts, imitation crabmeat, mayonnaise, chives, and dill; toss until equally combined. Mix in the cold pasta until evenly coated.

Steak Salad

Prep Time: 30 Minutes

Cook Time: 5 Minutes

Servings: 4

Ingredients:

- 3 teaspoons Worcestershire sauce
- ⅛ teaspoon liquid smoke flavoring
- 2 tablespoons lemon juice
- ½ teaspoon salt
- ⅛ teaspoon ground black pepper
- ½ cup sliced fresh mushrooms
- ¼ cup walnuts
- ⅓ cup vegetable oil
- 3 tablespoons red wine vinegar
- 1 ½ pounds beef sirloin steak
- 8 cups romaine lettuce, torn into bite-size pieces
- 6 roma (plum) tomatoes, sliced
- ¾ cup crumbled blue cheese

Instructions:

1. Preheat the oven to broil. Broil steaks for 3 to 5 minutes each side, or until done to preference. Allow to cool before slicing into bite-sized pieces.
2. Arrange lettuce, tomatoes, and mushrooms on cold plates. Top with blue cheese and walnuts. Serve with steak pieces on top.
3. Whisk together the oil, vinegar, lemon juice, salt, pepper, Worcestershire sauce, and smoke flavoring in a small bowl. Pour over salad.

Broccoli Mango Salad

Prep Time: 20 Minutes

Cook Time: 4 Hours

Servings: 8

Ingredients:

- 1 small red onion, cut into thin wedges
- 1 (11 ounce) can mandarin oranges, drained
- 2 tablespoons orange juice
- ½ cup reduced-fat ranch dressing
- 1 large mango, peeled and cubed
- ½ cup cashews
- 4 cups chopped broccoli
- 1 tablespoon prepared horseradish

Instructions:

1. In a mixing bowl, combine ranch dressing, orange juice, and horseradish; stir well and leave away.
2. In a salad bowl, combine the chopped broccoli, mango, cashews, and onion wedges. Toss the broccoli mixture with the dressing to coat. Refrigerate for at least four hours. Just before serving, add the drained oranges to the salad.

Bok Choy Salad

Prep Time: 20 Minutes

Cook Time: 20 Minutes

Servings: 4

Ingredients:

- ⅛ cup slivered almonds, toasted
- 3 tablespoons soy sauce
- ½ cup olive oil
- 2 bunches baby bok choy, cleaned and sliced
- 1 bunch green onions, chopped
- ½ (6 ounce) package chow mein noodles
- ¼ cup white vinegar
- ⅓ cup white sugar

Instructions:

1. Combine the olive oil, white vinegar, sugar, and soy sauce in a glass jar with a lid. Close the top and shake until well combined.
2. In a salad dish, combine the bok choy, green onions, almonds, and chow mein noodles. Toss with dressing before serving.

Beet and Feta Salad

Prep Time: 15Minutes

Cook Time: 45 Minutes

Servings: 4

Ingredients:

- 4 small beets, trimmed, leaving 1 inch of stems attached
- 4 cloves garlic, chopped, divided
- 3 heads baby bok choy, chopped
- ⅓ cup crumbled feta cheese
- 1 ½ teaspoons butter
- 2 tablespoons peanut oil
- 1 teaspoon olive oil

Instructions:

1. Preheat the oven to 425°F (220 degrees C). On a sheet of thick aluminum foil, place the beets, 1/4 of the chopped garlic, and the olive oil; wrap the foil over the beets to form a tight package.
2. Roast beets in a preheated oven for 40 minutes to 1 hour, or until readily punctured with a fork. Allow beets to cool just enough to handle, then wipe with a paper towel to remove skins. Set aside the 1/2-inch cubes.
3. In a big skillet over medium-high heat, heat the peanut oil and butter. Cook, stirring occasionally, until the bok choy is slightly softened but still crisp, approximately 5 minutes. Remove from the pan and mix in the beets and feta. Serve hot.

Mango-Prawn Salad

Prep Time: 35Minutes

Cook Time: 5 Minutes

Servings: 4

Ingredients:

- 2 tablespoons peanut oil
- 1-pound large shrimp, peeled and deveined
- 1 tablespoon chopped garlic
- 1 tablespoon fish sauce
- 1 red onion, thinly sliced
- ½ cup matchstick-cut carrots
- 1 ½ teaspoons chopped pickled ginger
- ½ cup mirin (Japanese sweet wine)
- ¼ cup chopped cashews
- ¼ cup chopped green onions, or more to taste
- ½ lime, juiced
- ½ teaspoon soy sauce, or more to taste
- 1 pinch white sugar
- 1 romaine hearts, chopped into bite-sized pieces, or more to taste
- 1 ripe mango, cubed
- 1 tablespoon rice vinegar
- ½ red pepper flakes, or more to taste

Instructions:

1. In a wok, heat the oil over medium-high to high heat. Cook until the shrimp begin to turn pink, approximately 2 minutes. Mix in the garlic and ginger. 1 minute in the oven Combine mirin, fish sauce, vinegar, and red pepper flakes in a mixing bowl. Combine the red onion, carrots, cashews, and green onions in a mixing bowl. 1 to 2 minutes of stirring Combine the mango, lime juice, soy sauce, and sugar in a mixing bowl. Remove the salad from the heat.
2. Arrange romaine hearts on a dish and spoon salad, including sauce, over top. Serve right away.

Chicken & Poultry

Teriyaki Chicken Sheet Pan Recipe

Prep Time: 10 Minutes

Cook Time: 40 Minutes

Servings: 4

Ingredients:

- 1 tbsp. grated ginger or ½ tsp dry
- 2 lbs chicken breasts – cut in half length ways to speed cooking
- 3 tsp minced garlic or 2 tsp dry
- 1 Tbsp. corn-starch
- ¼ cup water
- 1 Cup carrots sliced on diagonal
- 1 cup soy sauce
- 3 Cup broccoli florets
- 3 Tbsp. rice vinegar
- ¼ cup edamame shelled
- 3 Tbsp. honey
- 3 tsp sesame oil divided
- Sesame seeds for garnish optional

Instructions:

For the sauce:

1. Whisk together soy sauce, honey, vinegar, 2 tsp sesame oil, garlic, ginger, cornstarch, and water in a saucepan over medium heat. Simmer, stirring constantly, until the sauce thickens and bubbles. Set aside after removing from the heat.
2. For the chicken and veggies, prepare as follows:
3. Heat the oven to 400°F. Set aside a large sheet pan lined with parchment paper or foil sprayed with cooking spray.

4. Season each side of the chicken with salt and pepper, then pour about half of the sauce over the chicken, covering well on both sides.
5. Cook for 20 minutes in a preheated oven. Drizzle extra sauce over the chicken.
6. Arrange the broccoli florets, carrots, and edamame around the chicken in a single layer. Toss with 1 teaspoon sesame oil and 1 teaspoon teriyaki sauce to coat. Return to the oven for another 15-20 minutes, or until the chicken is well cooked and the juices run clear.
7. Remove the skillet from the oven, then slice the chicken into strips and sprinkle with the leftover sauce. Garnish with sesame seeds if preferred and serve over brown rice, quinoa, or cauliflower rice.

Chicken Quesadillas

Prep Time: 5 Minutes

Cook Time: 20 Minutes

Servings: 4

Ingredients:

- 2 to 3 teaspoons of fajita seasoning
- 2 chicken breasts
- 1 cup of low fat cheddar cheese or reduced fat Mexican Cheese Blend
- 4 Flour tortillas such as Carb Counter or Ole Extreme Welness Tortillas

Instructions:

1. Season the chicken with fajita spice.

2. Cook the chicken breasts in a skillet until they are done. When the chicken is no longer pink in the center and the internal temperature hits 165 degrees, it is done. If you're in a rush and don't want to prepare chicken at all, rotisserie chicken is an option.
3. Spray a nonstick pan with cooking spray and place one tortilla on it. 14 cup cheese and 14 cup chicken on top of the other tortilla
4. When the tortilla is golden and the cheese is melted, the quesadilla is done on one side. Wait until the other side is brown before flipping.

Chicken and Broccoli Dinner

Prep Time: 10 Minutes

Cook Time: 30 Minutes

Servings: 4

Ingredients:

- 12 to 18 ounces boneless and skinless chicken breasts chopped into large bite-sized pieces
- 1/2 cup yellow onion chopped
- 1/4 teaspoon kosher or sea salt
- 1/2 cup celery sliced
- 1/4 cup chicken broth or water
- 1 tablespoon extra virgin olive oil
- 2 cups broccoli florets
- 2 garlic cloves minced
- 1/4 teaspoon black pepper
- 2 tablespoons vegetable Sriracha
- 1/4 cup coconut aminos optional tamari or lite soy sauce

Instructions:

1. Brown the chicken on both sides in a pan over medium heat for approximately 8 minutes. Remove the chicken from the pan and put it aside. Sauté the broccoli florets until they are slightly soft. Remove and put aside the broccoli.
2. Cook for 5 to 8 minutes, until the celery has softened and the onion is translucent, in the pan with the onion. Cook for 30 seconds, or until the garlic is aromatic. Return the chicken, broccoli, and broth (or water) to the skillet, along with the salt and pepper. Cook for another 5 minutes, or until the chicken is well heated.
3. Whisk together the sauce ingredients and add to the pan for the final minute of cooking time. Serve and have fun!

Chicken Ranch Pasta

Prep Time: - Minutes

Cook Time: - Minutes

Servings: 6

Ingredients:

- 2 tablespoons of light butter
- ½ c fat free shredded cheddar cheese
- 8oz penne pasta
- 4 slice bacon, diced
- 1 large chicken breast, boneless & skinless, cut into bite sized pieces
- 1 tablespoon all-purpose flour
- ½ pkg dry ranch dressing mix
- 1 ½ cups fat-free milk
- 2 scallions, chopped

- salt & pepper to taste

Instructions:

1. Cook pasta in boiling water according to package instructions; drain and put aside.
2. Meanwhile, fry the bacon in a large pan over medium heat. Dry with paper towels. Remove any excess bacon grease from the pan.
3. Season the chicken well with salt and pepper. In the same skillet, melt the butter and add the chicken. Cook until the potatoes are soft, no longer pink, and slightly browned.
4. Stir the flour and ranch dressing mix into the chicken to coat evenly. Cook, stirring periodically, until the sauce is thickened and bubbling. Cook and stir until half of the cheddar cheese and half of the saved bacon are melted.
5. Season with salt and pepper to taste.
6. Serve each serving of spaghetti with more bacon sprinkled on top. Garnish with green onions if desired (optional)
7. 1 cup serving size yields 6 servings.

Costa Brava Chicken

Prep Time: 5 Minutes

Cook Time: 25 Minutes

Servings: 4

Ingredients:

- 1 tablespoon vegetable oil
- 1 teaspoon cumin, grounded
- 1 onion, quartered

- ½ cup salsa
- 2 tablespoons water
- 1 red bell pepper, thinly sliced
- 10 chicken breast halves, skinless and boneless
- 2 garlic cloves, minced
- 1 can (14 ounces) stewed tomatoes
- 1 can (20 ounces) pineapple chunks
- 2 cups black olive
- 1 teaspoon cinnamon, grounded
- Salt as needed

Instructions:

1. Take a pineapple and thoroughly wash it.
2. Drain the pineapples and keep the juice aside.
3. Season with salt to taste.
4. In a big frying pan, heat the oil.
5. Place it over medium heat and sprinkle with cumin and cinnamon.
6. Mix in the garlic and onion.
7. Cook until the pineapple is soft. Stir in the salsa, olives, and tomatoes.
8. Put the cover back on.
9. After that, simmer for 25 minutes.
10. Fill a basin halfway with cornstarch and water.
11. Stir the ingredients into a skillet. Add the bell pepper and continue to cook until the sauce boils and thickens.
12. Mix in the pineapple pieces.
13. Serve immediately and enjoy!

Mango Flavored Chicken

Prep Time: 25 Minutes

Cook Time: 10 Minutes

Servings: 4

Ingredients:

- 14-ounce chicken breast halves cut up into skinless and boneless
- cubes
- 4 medium-sized shallots
- 4 teaspoon vegetable oil
- 1 large-sized English cucumber, seeded and sliced
- 10-ounce coconut milk
- 2 medium-sized mangoes, peeled and sliced
- 4 teaspoon spicy curry paste

Instructions:

1. Put half of the mangoes in a bowl and add the halves to it.
2. In a blender, combine mangoes and coconut milk.
3. Blend until the purée is smooth.
4. Place the mixture to the side.
5. Place a big saucepan on the stove over medium heat.
6. Heat up the oil
7. Cook for 1 minute after adding the curry paste.
8. Cook for 5 minutes with the shallots and chicken.
9. Allow the mango puree to cook up in the mixture.
10. Cooked chicken should be accompanied with mango puree and cucumbers.
11. Serve immediately and enjoy!

Orange Chicken

Prep Time: 20 Minutes

Cook Time: 7 Minutes

Servings: 4

Ingredients:

- orange juice from 1 whole orange
- orange zest from 1 whole orange
- 1 tbsp ginger I prefer fresh :)
- 1 tsp crushed red peppers
- 1 lbs chicken breast cut into 1 inch cubes
- 1/4 cup coconut aminos or low sodium tamari or soy sauce
- 1/4 cup vinegar rice vinegar
- 1/4 cup chicken broth low sodium
- 6 tsp honey
- 2 cloves minced garlic
- 2 tsp arrowroot powder or corn starch

Instructions:

1. Cut your chicken into 1 inch bite-sized chunks.
2. Set your Instant Pot insert to sauté and lightly spray with oil.
3. In a mixing bowl, combine the chicken broth, soy sauce, rice vinegar, honey, ginger, crushed red pepper, garlic, orange zest, and orange juice.
4. Once the sauce is boiling, add the cubed chicken and toss until evenly covered with the sauce.
5. Close and lock the lid, then turn the valve to the sealing position. Choose Pressure Set the timer for 7 minutes and start cooking.

6. When the cooking time is up, let the pressure to freely release before removing the cover.
7. Transfer 12 cup of the sauce to a small bowl and stir in the arrowroot flour. Return the sauce to the saucepan and stir thoroughly. Allow to sit until the sauce thickens.
8. If preferred, top with more ginger and red pepper and serve over brown rice or brown rice noodles.

Juicy Chicken Curry

Prep Time: 10 Minutes

Cook Time: 20 Minutes

Servings: 4

Ingredients:

- 3 boneless chicken breasts, cut into 1-inch portions
- 2/3 cup chicken broth
- 3 tablespoon fish sauce
- 1 can (8 ounces) bamboo shoots, sliced and drained
- 1-piece green bell pepper, cut into 1-inch pieces
- ¼ cup fresh basil, chopped
- 1 cup fresh mushrooms, sliced
- 2 cans coconut milk
- 1 can (8 ounces) water chestnuts, sliced and drained
- 2 tablespoons green curry paste

Instructions:

1. Heat the coconut milk and curry paste in a pan over medium heat.
2. Simmer for 5 minutes, whisking constantly.

3. Add the chicken, chicken broth, water chestnuts, bamboo shoots, mushroom, and bell pepper and mix well.
4. Season to taste with fish sauce.
5. Stir in the basil.
6. Cook for 10 minutes.
7. Serve and have fun!

Slowly Cooked Chicken Chili

Prep Time: 10 Minutes

Cook Time: 8 Hours

Servings: 8

Ingredients:

- 1 3/8-ounce chili seasoning mix
- 15-ounce corn, undrained
- 15-ounce kidney beans, drained
- 12-ounce chicken breast, diced, skinless, and boneless
- 1 green bell pepper, chopped
- 1 onion, chopped
- 1-quart tomatoes, canned
- ½ cup salsa

Instructions:

1. Grease a pan generously with cooking spray.
2. Transfer the chicken to the pan and brown it.
3. In the Slow Cooker, combine the chicken, chili spice mix, tomato, corn, kidney beans, green bell pepper, salsa, and onion.
4. Close the lid.

5. Serve and have fun!

WW Chicken Chili

Prep Time: 5 Minutes

Cook Time: 20 Minutes

Servings: 6

Ingredients:

- 1 can green chilis
- 1 can 15 oz pinto beans
- 1 can 16 oz. kidney beans'1 can diced tomatoes
- 2 cups cooked chicken shredded
- ½ cup chopped onions
- 1 teaspoon chili powder
- ½ cup red salsa
- 1 teaspoon ground cumin
- ¼ teaspoon pepper
- 1 teaspoon salt
- 2 cups chicken stock
- ½ cup water
- 1 clove minced garlic

Instructions:

1. In a medium-high-heat saucepan, combine all of the ingredients.
2. Bring to a boil.
3. Reduce heat to medium-low and cook for 20 minutes (covered).

Baked Lemon Chicken and Herb

Prep Time: 10 Minutes

Cook Time: 35 Minutes

Servings: 4

Ingredients:

- ¼ cup chicken broth
- 1 medium-sized quartered lemon
- 2 teaspoons fresh lemon juice
- ½ teaspoon sea salt
- 1 teaspoon olive oil
- 2 teaspoons fresh parsley, chopped
- 1-pound chicken breast, skinless and boneless
- 2 teaspoons rosemary, chopped
- ¼ teaspoon black pepper, freshly ground

Instructions:

1. Preheat your oven to 400 degrees Fahrenheit.
2. Season the chicken on both sides with sea salt and black pepper.
3. Use a small shallow roasting pan.
4. Drizzle with olive oil and sprinkle with lemon juice, rosemary, and parsley.
5. Coat the chicken with the broth and bake for 30-35 minutes.
6. Garnish with lemon slices
7. Serve and have fun!

Lemony Chicken Tenders

Prep Time: 10 Minutes

Cook Time: 25 Minutes

Servings: 4

Ingredients:

- cooking spray
- ¼ teaspoon ground paprika
- 1 (16 ounce) package boneless, skinless chicken tenders
- ½ teaspoon salt
- ⅓ cup lemon juice
- 4 teaspoons unsalted butter, softened

Instructions:

1. Preheat the oven to 425° F. (220 degrees C). Coat an 8x11-inch baking pan with nonstick cooking spray.
2. In the preheated pan, arrange the chicken tenders in a single layer. Spread butter on the tenders, then season with salt and paprika to taste. Drizzle with lemon juice to finish.
3. Bake uncovered in a preheated oven for 25 to 30 minutes, or until the chicken is no longer pink in the middle. In the middle, an instant-read thermometer should read at least 165 degrees F. (74 degrees C).

Salsa Chicken Meal

Prep Time: 5 Minutes

Cook Time: 40 Minutes

Servings: 4

Ingredients:

- ¼ teaspoon dried thyme
- 1 (14.5 ounce) can diced tomatoes, drained
- 3 cups tomato-vegetable juice cocktail
- ¾ cup chopped onion
- 1 cup chopped carrots
- 1 potato, peeled and cubed
- 2 (15 ounce) cans garbanzo beans, drained
- 1 cup long-grain white rice
- 1 tablespoon soy sauce
- 1 leek, chopped
- ½ teaspoon ground black pepper
- ¼ teaspoon garlic powder
- 1 (15 ounce) can peas, drained
- 2 cups water
- ¾ cup chopped celery
- 1 teaspoon dried dill weed
- 1 (15 ounce) can whole kernel corn, drained

Instructions:

1. In a large saucepan over high heat, combine celery, onion, carrots, tomatoes, tomato-vegetable juice, water, leek, potato, peas, corn, garbanzo beans, rice, soy sauce, thyme, crushed black pepper, garlic powder, and dill weed.

2. Bring to a boil, then lower to a low heat and cook until the veggies are soft, approximately 30 minutes.

Parmesan Chicken Cutlets

Prep Time: 15 Minutes

Cook Time: 20 Minutes

Servings: 4

Ingredients:

- 1⁄2 teaspoon garlic powder
- 4 (1/4 lb) boneless skinless chicken breast
- 1⁄4 teaspoon fresh ground pepper
- 1⁄4 cup parmesan cheese, grated
- 1teaspoon dried parsley
- 1⁄8 teaspoon paprika
- 2 tablespoons dried Italian seasoned breadcrumbs

Instructions:

1. Preheat the oven to 400°F.
2. Combine cheese, crumbs, and all spices in a resealable plastic bag; shake thoroughly.
3. Place the cheese mixture on a platter and dip each chicken breast in it, turning to cover both sides.
4. Place on a nonstick baking sheet.
5. Bake for 20-25 minutes, or until the chicken is cooked through.

Meat Recipes

Lamb Cutlets

Prep Time: 10 Minutes

Cook Time: 10 Minutes

Servings: 4

Ingredients:

- 480g lamb cutlet, trimmed and fat-free
- 1 tablespoon honey
- 1 tablespoon Dijon mustard
- 250g brown quick rice, cooked
- 2 cups baby spinach leaves, finely shredded
- 2 tablespoon white wine vinegar

Instructions:

1. In a small mixing bowl, combine the mustard, honey, and vinegar. Preheat your grill to high and cook the cutlets for approximately 2 minutes each side.
2. Place the rice in a microwave-safe bowl.
3. Place them in the microwave to cook.
4. While the rice is heating up, stir in the spinach leaves. Serve with the grilled cutlets.
5. Drizzle the remaining mustard mixture over top and serve immediately.

Slow Cooker Pulled Pork

Prep Time: 15 Minutes

Cook Time: 5 Hours

Servings: 8

Ingredients:

- ½ cup apple cider vinegar
- 1 tablespoon prepared yellow mustard
- 1 teaspoon vegetable oil
- 2 large cloves garlic, crushed
- 2 tablespoons butter, or as needed
- 1 ½ teaspoons dried thyme
- 8 hamburger buns, split
- 1 (4 pound) pork shoulder roast
- 1 cup barbeque sauce
- ½ cup chicken broth
- 1 tablespoon Worcestershire sauce
- 1 tablespoon chili powder
- 1 extra large onion, chopped
- ¼ cup light brown sugar

Instructions:

1. Fill the slow cooker halfway with vegetable oil. Pour the barbecue sauce, vinegar, and chicken stock into the slow cooker with the pork roast. Brown sugar, yellow mustard, Worcestershire sauce, chili powder, onion, garlic, and thyme are all good additions. Cook for 10 to 12 hours on Low or 5 to 6 hours on High, or until pork shreds easily with a fork.

2. Remove the pork from the slow cooker and shred it with two forks. Stir the shredded pork back into the slow cooker with the liquids.
3. Spread butter on the insides of both hamburger bread halves. Toast the buns in a pan, butter side down, over medium heat until golden brown. Fill toasted buns with pulled pork.

Southern Pork Delight

Prep Time: 5 Minutes

Cook Time: 10 Minutes

Servings: 8

Ingredients:

- 2 ½ cups milk
- Salt and ground black pepper to taste
- 1 (16 ounce) can refrigerated jumbo buttermilk biscuits
- ¼ cup flour
- 1 (9.6 ounce) package Jimmy Dean® Original Hearty Pork Sausage Crumbles

Instructions:

1. Bake the biscuits according to the package instructions.
2. Meanwhile, in a large pan over medium heat, sauté sausage for 5-6 minutes, or until well cooked, turning constantly. Mix in the flour. Cook, stirring regularly, until the liquid comes to a boil and thickens. Reduce the heat to medium-low and continue to whisk regularly for 2 minutes. Season with salt and pepper to taste.
3. Cut the biscuits in half. Top each dish with 2 halves and roughly 1/3 cup gravy.

Maple Roasted Brussels Sprouts with Bacon

Prep Time: 10 Minutes

Cook Time: 20 Minutes

Servings: 6

Ingredients:

- ¼ cup extra-virgin olive oil
- ½ teaspoon salt
- 1 pound whole Brussels sprouts
- 3 tablespoons pure maple syrup
- 4 slices bacon, cut into 1/2-inch pieces
- ¼ teaspoon freshly ground black pepper

Instructions:

1. Preheat the oven to 400° F. (200 degrees C). Aluminum foil should be used to line a rimmed baking pan.
2. Trim the ends of any big Brussels sprouts and chop them in half. Transfer to a large mixing bowl.
3. Combine the Brussels sprouts with the bacon, salt, and pepper. Drizzle olive oil and maple syrup over top and toss to coat sprouts. Spread in a single layer on the prepared baking sheet.
4. Cook until the bacon is crispy and the Brussels sprouts are caramelized, 20 to 30 minutes, stirring halfway through.

Onion And Pork Cutlets

Prep Time: 5 Minutes

Cook Time: 2 Hours 20 Minutes

Servings: 20

Ingredients:

- 2 pork cutlets
- 1 tablespoon olive oil
- ½ teaspoon paprika
- 1 Spanish onion, chopped
- Salt and pepper to taste
- 1 bell pepper, deveined and sliced
- 2 garlic cloves, minced
- ½ teaspoon hot sauce
- ½ teaspoon mustard

Instructions:

1. Preheat the oven to 325° F. (165 degrees C). In a roasting pan, place the ham.
2. Bake the ham for 2 hours, uncovered, in a preheated oven.
3. In a saucepan, mix brown sugar, pineapple juice, and 1/3 cup honey while the ham is baking. Squeeze the orange juice into the pot after zesting it. Whisk in the Dijon mustard and ground cloves as the sauce comes to a boil over medium-high heat. Reduce heat to low and continue to cook until the glaze thickens slightly, 5 to 10 minutes.
4. Season with salt and pepper to taste. Add lemon juice for additional sharpness, or honey for sweetness. Place aside.

5. After 2 hours of baking, remove the ham from the oven. Glaze using a brush. Bake for 30 to 45 minutes more, coating the ham with glaze every 10 minutes.

Oven-Baked Baby Back Ribs

Prep Time: 15 Minutes

Cook Time: 2 Hours 30 Minutes

Servings: 4

Ingredients:

- ½ cup ancho chile powder
- ½ teaspoon ground dried chipotle pepper
- ¼ cup brown sugar
- ¼ cup salt
- ¼ cup white sugar
- 1 tablespoon ground cumin
- 1 teaspoon ground cayenne pepper
- 1 teaspoon dry mustard
- 1 cup barbeque sauce
- 2 tablespoons freshly ground black pepper
- 1 rack baby back pork ribs

Instructions:

1. Preheat the oven to 250° F. (120 degrees C).
2. In a small mixing bowl, add ancho chile powder, white and brown sugars, salt, black pepper, cumin, dry mustard, cayenne pepper, and chipotle pepper.

3. Place the rib rack on heavy-duty aluminum foil, meat-side down. Using a knife, prick the back of the rack many times.
4. Apply a generous layer of dry rub on all sides of the rack. Keep any leftover dry rub in an airtight jar for later use.
5. Fold the foil around the rack, meat side down, to make a tight seal. Place on a baking sheet.
6. Bake until soft and cooked through, approximately 2 hours in a preheated oven. Remove and set aside for 15 minutes to cool.
7. Raise the oven temperature to 350°F (175 degrees C).
8. Open the foil and drain any collected juices and fat. Brush the rack with barbecue sauce on both sides.
9. Place the rack on the foil with the meat facing up. Bake for 10 minutes with the foil still open in the oven. Remove from the oven and coat the meat side with another layer of barbecue sauce. Repeat baking and coating with sauce four more times for a total baking duration of 50 minutes.
10. Serve the rack in separate rib slices with more barbecue sauce.

Awesome BBQ Loaf

Prep Time: 5 Minutes

Cook Time: 30 Minutes

Servings: 4

Ingredients:

- ¼ cup of seasoned Italian bread crumbs
- ½ a cup of BBQ sauce
- ¼ cup of frozen chopped onion pressed dry
- 1 pound 93% lean ground beef

- 2 large egg whites

Instructions:

1. Preheat your oven to 375 degrees F.
2. In a mixing bowl, combine the meat, 14 cup of BBQ sauce, onion, egg whites, bread crumbs, and spice of choice.
3. Stir
4. Form the batter into a loaf pan.
5. Spread the remaining 14 cup BBQ sauce over the loaf.
6. Bake at 375 degrees Fahrenheit for 40 minutes.
7. Check for doneness and have fun!

Slow Roast Leg of Lamb

Prep Time: 15 Minutes

Cook Time: 5 Hours 15 Minutes

Servings: 3

Ingredients:

- 2.25 kg / 4.5 lb leg of lamb, bone in
- 1.5 tbsp olive oil
- 1 whole garlic head, unpeeled, cut in half horizontally
- 3 cups beef stock/broth, low sodium
- Salt and pepper to taste
- 2 rosemary sprigs
- 1 onion, quartered
- 2 cups water
- 1 cup of water
- Salt and pepper, to taste
- 4 tbsp. flour

Instructions:

1. Preheat the oven to 170°C/335°F (standard) or 150°C (lower) (fan).
2. In a metal roasting pan, combine the garlic, onion, and rosemary.
3. Season lamb as follows: Place the lamb leg in the pan, right side up. (Note 2) Season liberally with salt and pepper and massage in.
4. Turn the lamb over and set it on top of the garlic and onion. Rub in extra salt and pepper if desired. (A video is useful here.)
5. Cover and add liquids: Drizzle olive oil over the lamb. Pour broth and water around the lamb - it won't cover it, but that's OK since the meat will sink into it. Cover with foil (don't use a lid; you want some liquid to escape).
6. Place in the oven for 4.5 hours to slow roast. (For a roasting time chart, see Notes.)
7. Check the meat: Remove from the oven and peel off the foil. Turn the lamb over. Check to see whether the flesh is soft enough to easily take a piece off with a fork. If not, return to the oven, covered.
8. Return the uncovered lamb to the oven for another 45 minutes, or until nicely browned.
9. Rest the lamb by spooning pan juices over it. Transfer to a serving tray and loosely cover with foil while you prepare the gravy (stays warm for 1 – 1.5 hours).

GRAVY:

1. Skim fat: Using a big spoon, remove and discard part of the fat off the liquid's surface.
2. Place the pan on medium high heat and add the flour. When the liquid begins to bubble, add the flour. Mix it in with a whisk; this may take a few minutes as the liquid decreases.
3. Add water: Once it looks like sludge (see video), whisk in 1/2 - 1 cup of water until it reaches the desired gravy consistency. Season with salt and pepper to taste; I seldom use more salt.

4. Strain the gravy into a basin, pressing the onion juices out. Pour the gravy into the jug.

Spaghetti Sauce with Ground Beef

Prep Time: 10 Minutes

Cook Time: 1 Hours 15 Minutes

Servings: 8

Ingredients:

- 2 teaspoons dried oregano
- 1-pound ground beef
- 1 (6 ounce) can tomato paste
- 4 cloves garlic, minced
- ½ teaspoon ground black pepper
- 1 (16 ounce) can tomato sauce
- 1 (28 ounce) can diced tomatoes
- 1 teaspoon salt
- 2 teaspoons dried basil
- 1 small green bell pepper, diced
- 1 medium onion, chopped

Instructions:

1. In a large saucepan over medium-high heat, combine ground beef, onion, garlic, and green pepper. Cook and stir for 5 to 7 minutes, or until the beef is browned and crumbly and the veggies are soft. Grease should be drained.

2. In the pan, combine chopped tomatoes, tomato sauce, and tomato paste. Oregano, basil, salt, and pepper to taste. 1 hour, stirring periodically, simmer spaghetti sauce

Slow-Cooker Corned Beef and Cabbage

Prep Time: 15 Minutes

Cook Time: 8 Minutes

Servings: 8

Ingredients:

- 10 baby red potatoes, quartered
- 1 onion, peeled and cut into bite-sized pieces
- 6 ounces' beer
- 1 (4 pound) corned beef brisket with spice packet
- 4 large carrots, peeled and cut into matchstick pieces
- 4 cups water
- ½ head cabbage, coarsely chopped

Instructions:

1. Fill the bottom of a slow cooker with potatoes, carrots, and onion; add water and top with beef. Pour the beer over the brisket, then sprinkle with the spices from the package and cover.
2. Simmer for 7 hours on High, then add the cabbage and cook for another hour.

Cast Iron Pan-Seared Steak

Prep Time: 5 Minutes

Cook Time: 13 Minutes

Servings: 2

Ingredients:

- freshly ground black pepper to taste
- 2 teaspoons olive oil
- 1 cup apple cider vinegar
- 2 (6 oz.) beef top sirloin steaks
- 2 cups of orange juice
- 1 ½ tablespoons steak seasoning, or to taste
- ½ cup Worcestershire sauce
- sea salt to the taste

Instructions:

1. Place the steaks in a large casserole dish, side by side. Combine the orange juice, cider vinegar, and Worcestershire sauce in a mixing bowl. Marinate for 45 minutes in the refrigerator, uncovered.
2. Preheat the oven to 425° F. (220 degrees C).
3. Take the casserole dish out of the refrigerator. Wrap the steaks in plastic wrap and let aside for at least 15 minutes to come to room temperature.
4. Rub steaks liberally with steak spice and black pepper on a clean work surface.
5. In a cast-iron skillet, heat the olive oil over high heat.

6. Cook the steaks in the heated pan for 2 1/2 minutes, or until lightly browned on the bottom. Cook until the opposite side is browned and the center is red, approximately 2 minutes longer. Place the skillet with the steaks in the oven.
7. 8 to 10 minutes in a preheated oven, until steaks are firm and reddish-pink to faintly pink in the middle. An instant-read thermometer put into the middle should read between 130° F (54° C) and 140° F. (60 degrees C).
8. Remove the steaks from the oven and season with salt. Allow for a 5-minute pause before serving.

Cuban Pork Chops

Prep Time: 5 Minutes

Cook Time: 15 Minutes

Servings: 5

Ingredients:

- 1 Small Uncooked Red Onions Thinly Sliced
- 1/4 Cup Cabelas Tequila Lime
- 2 Tbsp. Cilantro Fresh
- 2 Tsp Ground Cumin
- 4 Tsp Olive Oil
- ¼ Tsp Salt
- 6 Tbsp. Lime Juice
- ½ tsp Ground black pepper
- 3 Tbsp. Lime Juice
- 1 Tsp Salt

- 2 Tsp Oregano
- 3 Tsp Minced Garlic
- 1 Lb Boneless Pork Chops

Instructions:

1. Begin by cutting your onion and cilantro; the quantity of each may be adjusted to suit your family's tastes.
2. Combine the topping ingredients in a mixing basin and set aside for 30 minutes. (I kept mine in the fridge while the beef marinated.)
3. Place all marinate ingredients (except the pork chops) in a gallon Ziploc bag.
4. Shake or push the bag around to combine the marinated ingredients before adding the pork chops.
5. Allow the pork to marinate for as long as you want; mine took around 45 minutes, but you could marinade it overnight.
6. Warm up your pan or grill with oil or grease before cooking your pork chops.
7. Allow the pork chops to cook and turn them to ensure perfect cooking.
8. Plate the pork chops and top with marinated onions for more taste.
9. Serve with mashed potatoes or salad.

Seafood Recipes

Seafood Linguine

Prep Time: 10 Minutes

Cook Time: 30 Minutes

Servings: 4

Ingredients:

- 1/2 lb medium shrimp, cooked and peeled
- 1/2 lb crabmeat, cooked, picked over and flaked
- 1/4 teaspoon dried thyme leaves
- 1 tablespoon fresh lemon juice
- 4 cups cooked linguine
- 1/4 teaspoon crushed red pepper flakes
- 3 garlic cloves, minced
- 8 scallions, thinly sliced
- 1 tablespoon extra-virgin olive oil
- 2 tablespoons dry white wine
- 1/4 teaspoon salt

Instructions:

1. Heat the oil in a big nonstick skillet.
2. Cook, stirring regularly, until the scallions and garlic are softened, about 2-3 minutes.
3. Cook, stirring regularly, until the shrimp, crab, wine, lemon juice, pepper flakes, thyme, and salt are cooked through, approximately 3 minutes.
4. Serve the linguine with the seafood mixture on top.

Creamy Shrimp and Bacon

Prep Time: 20 Minutes

Cook Time: 8 Minutes

Servings: 3

Ingredients:

- ¼ cup coconut cream
- 1-pound raw shrimp
- 1 tablespoon olive oil
- 2 bacon slices
- Salt and black pepper, to taste

Instructions:

1. Season the shrimp with black pepper and salt.
2. In a nonstick skillet, heat the olive oil and seasoned shrimp.
3. After 5 minutes, add the bacon and coconut cream.
4. Cook for approximately 8 minutes before serving.

Garlic Butter Baked Salmon

Prep Time: 10 Minutes

Cook Time: 25 Minutes

Servings: 4

Ingredients:

- 2 tablespoons fresh chopped parsley
- 1/3 cup freshly squeezed lemon juice
- 1/2 cup melted unsalted butter

- 2 tablespoons dry white wine
- 1 lemon sliced to garnish
- 4 (6 oz | 170 g) skinless salmon fillets
- 2 1/2 tablespoons minced garlic, divided
- 2 tablespoons olive oil
- 1 pound (500 g) fingerling potatoes, halved
- 1 1/2 teaspoons salt, divided
- 1/2 teaspoon cracked black pepper, divided
- 3 bunches asparagus

Instructions:

1. Preheat the oven to 400°F | 200°C. Toss the potatoes with the oil, 1/2 tablespoon garlic, 1/2 teaspoon salt, and 1/4 teaspoon pepper on a large rimmed baking sheet. Spread them out evenly and roast for 15 minutes, or until they begin to soften and brown.
2. Arrange the potatoes on one side of the sheet pan and the salmon along the center. 1 1/2 teaspoons minced garlic and 2 tablespoons parsley rubbed evenly over salmon Place the asparagus on the other side of the pan.
3. Combine 1/4 cup lemon juice and 1/4 cup melted butter in a mixing bowl and pour over the fish and asparagus. Season with the remaining salt and pepper to taste.
4. Return to the oven and bake until the potatoes are golden and fork-tender and the salmon is opaque all the way through (about 10 minutes). Broil for 2 minutes towards the end if you want burnt edges.
5. Meanwhile, mix the remaining butter, garlic, and lemon juice with the wine in a small dish (or chicken stock). Serve with salmon, vegetables, and lemon slices!

Lemon Garlic Shrimp

Prep Time: 20 Minutes

Cook Time: 15 Minutes

Servings: 3

Ingredients:

- 4 garlic cloves
- 2 lemons, sliced
- 1 lb. large raw shrimp
- 3 tablespoons butter
- 1 teaspoon paprika

Instructions:

1. In a pan, melt the butter and add the garlic.
2. After 1 minute, add the shrimp, paprika, and lemon slices.
3. Cook for roughly 10 minutes on medium low heat before serving.

Spicy Shrimp and Quinoa Bowl

Prep Time: 5 Minutes

Cook Time: 8 Minutes

Servings: 2

Ingredients:

- 6 ounces' shrimp, peeled and deveined
- 2 tablespoons olive oil
- 1 teaspoon garlic powder

- 1 teaspoon dried oregano
- 2-½ cups sliced zucchini
- ½ teaspoon red pepper flakes
- ¼ teaspoon salt
- 1 cup grape tomatoes, halved
- 1 cup cooked quinoa

Instructions:

1. Warm the oil in a skillet over medium heat. Cook for 3-4 minutes after adding the zucchini and tomatoes.
2. Sprinkle the shrimp with garlic powder, salt, red pepper flakes, and oregano. Cook the shrimp for 3-4 minutes, or until they become pink.
3. Serve with quinoa.

Creamy Lemon Parmesan Shrimp

Prep Time: 10 Minutes

Cook Time: 15 Minutes

Servings: 4

Ingredients:

- 1 tablespoon butter
- 2 tablespoons minced garlic
- 1 pound | 500 grams jumbo-sized shrimp
- 3-4 tablespoons lemon juice
- 1 teaspoon corn-starch (cornflour) mixed with 1 tablespoon of milk
- 1 1/2 cups milk
- 1/3 cup fresh grated Parmesan cheese

Instructions:

1. In a large skillet, melt the butter (or oil) over medium-high heat until the butter is melted and the pan is hot. Fry the shrimp (prawns) for 2 minutes on each side, or until they are pink and 'just' done. Place on a platter.
2. Cook until the garlic is fragrant in the remaining butter (or oil) in the pan (about 1 minute). Reduce the heat to low-medium and stir in the milk (or cream). Bring the sauce to a boil; season with salt and pepper to taste; add the parmesan cheese and continue to cook for approximately 2 minutes, or until the sauce thickens. Pour the cornstarch/water mixture into the center of the pan and stir quickly to incorporate into the sauce. It will instantly begin to thicken.
3. Remove from the heat and mix in the lemon juice. Return the shrimp (prawns) to the pan and serve with lemon slices, parsley, more parmesan cheese, cracked pepper, and salt (if desired).
4. Over steamed veggies, zucchini noodles, rice, or cooked pasta, serve.

Cheesy and Creamy Tilapia

Prep Time: 10 Minutes

Cook Time: 30 Minutes

Servings: 4

Ingredients:

- 4 tilapia fillets
- 1 cup Parmesan cheese, grated
- ¼ cup fresh lemon juice
- ¼ cup mayonnaise
- Salt and black pepper, to the taste

Instructions:

1. Preheat the oven to 350°F and lightly butter two baking trays.
2. Marinate the tilapia fillets in mayonnaise, fresh lemon juice, salt, and pepper.
3. Place the marinated fillets in baking trays and sprinkle with cheese.
4. Preheat the oven to 350°F and bake for 30 minutes.
5. Remove from the oven and serve immediately.

Salmon and Avocado Rye

Prep Time: 10 Minutes

Cook Time: 0 Minutes

Servings: 2

Ingredients:

- 2 teaspoons lime juice
- 1 tablespoon fresh chives, chopped
- 5-ounce salmon, smoked
- 2 slices dark rye bread
- 1 medium fresh Lebanese cucumber, thinly sliced
- ½ tsp chili flakes
- ½ medium avocados
- ¾ oz. Trimmed snow peas

Instructions:

1. In a small bowl, combine the avocados, chives, juice, and chili.
2. Using a fork, mix everything together.
3. Create a smooth mixture.

4. Season with salt and freshly ground black pepper.
5. Toast the bread
6. Spread the avocado mixture on the bread.
7. Serve with salmon, cucumber, and snow pea shoots.
8. Enjoy!

Tuna Patties

Prep Time: 10 Minutes

Cook Time: 8 Minutes

Servings: 4

Ingredients:

- 1 raw egg
- 2 tablespoons chopped fresh chives, green onions, or shallots
- Salt and freshly ground black pepper, to taste
- Crystal or Tabasco hot sauce, to taste
- 2 (5 to 6-ounce) cans tuna
- 1 teaspoon lemon zest
- 1 tablespoon lemon juice
- 2 teaspoons Dijon mustard
- 1/2 cup white bread, torn into small pieces
- Lemon wedges, for serving
- 2 tablespoons extra virgin olive oil
- 1/2 teaspoon butter
- 1 tablespoon water, or liquid from the cans of tuna
- 2 tablespoons chopped fresh parsley

Instructions:

1. Drain the tuna: Remove the tuna cans' liquid. If using canned tuna, save aside a tablespoon of the tuna water and add a teaspoon of olive oil to the tuna mixture in the following step.
2. Make the tuna mixture as follows:
3. Combine the tuna, mustard, torn white bread, lemon zest, lemon juice, water, parsley, chives, and spicy sauce in a medium mixing basin. Season with salt and freshly ground black pepper to taste. Before adding the egg, taste the mixture to check if it needs extra spice to your liking. Incorporate the egg.
4. Form into patties and set aside to cool:
5. Divide the mixture into four equal pieces. Form each portion into a ball, then flatten into a patty. Chill for one hour on a tray lined with wax paper. (If you are short on time, you may skip the chilling.) Chilling just helps the patties hold together when cooking.)
6. In a pan, heat the olive oil and a little butter (to taste) on medium high in a cast iron or nonstick skillet. Place the patties in the skillet and cook for 3 to 4 minutes on each side, or until beautifully browned.
7. Serve with lemon wedges. Tuna burgers may also be served with tartar sauce on slider buns.

Grilled Shrimp Skewers

Prep Time: 20 Minutes

Cook Time: 15 Minutes

Servings: 6

Ingredients:

- 2 tbsp. olive oil

- 1 tsp. minced garlic
- 1 tsp. smoked paprika
- 1 tbsp. lemon juice
- Red pepper flakes to taste
- 1 lb. peeled cleaned shrimp
- Sea salt flakes to taste

Instructions:

1. Before usage, soak bamboo skewers in water for 30 minutes.
2. In a mixing bowl, combine olive oil, lemon juice, smoked paprika, garlic, red pepper flakes, and salt.
3. Mix in the shrimp until they are evenly coated.
4. Allow the shrimp to rest for at least 30 minutes.
5. Preheat the grill (or charcoal grill) to medium heat, brush the grates with oil, and cook the shrimp for 2-3 minutes each side, or until opaque.
6. Transfer the skewers to a dish and serve.

Garlic Lemon Scallops

Prep Time: 10 Minutes

Cook Time: 5 Minutes

Servings: 4

Ingredients:

- 1 and ¼ pound scallops, dried
- 2 tablespoons all-purpose flour
- 1 pinch ground sage

- 2 tablespoon parsley, chopped
- 1 tablespoon olive oil
- 1 scallion, finely chopped
- 1 juice lemon
- ¼ teaspoon salt
- 4-6 pieces' garlic cloves, minced

Instructions:

1. Pour some oil into a big nonstick skillet.
2. Warm it up
3. Take a medium-sized mixing basin.
4. Season the scallops with salt and pepper.
5. Fill your skillets with scallops, scallions, garlic, and sage.
6. Cook the scallops for 3-4 minutes, or until they are opaque.
7. Combine the lemon juice and parsley in a mixing bowl.
8. Turn off the heat.
9. Serve right away and enjoy!

Salmon Stew

Prep Time: 10 Minutes

Cook Time: 5 Hours

Servings: 9

Ingredients:

- 3 pounds' salmon fillet, cubed
- 3 cups homemade fish broth3 tablespoons butter

- 3 medium onions, chopped
- Salt and black pepper, to the taste

Instructions:

1. In a slow cooker, combine all of the ingredients and stir gently.
2. Cook for roughly 5 hours on Low in the slow cooker.
3. Remove from the oven and serve immediately.

Sweet and Sour Fish

Prep Time: 7 Minutes

Cook Time: 20 Minutes

Servings: 6

Ingredients:

- 2 pounds' fish chunks
- ¼ cup butter
- 2 tablespoons vinegar
- 4 drops liquid stevia
- Salt and black pepper, to taste

Instructions:

1. In a pan, heat the butter and fish for around 4 minutes.
2. Add the stevia, vinegar, salt, and black pepper to taste.
3. Cook for roughly 10 minutes on medium low heat before serving.

Fluffy Haddock and Potato Pie

Prep Time: 20 Minutes

Cook Time: 40 Minutes

Servings: 4

Ingredients:

- 1 ½ pounds haddock fillets
- 1 ¼ cups milk
- 1 cup creme fraiche
- 3 ounces Emmentaler cheese, finely shredded
- 3 ½ pounds potatoes, peeled and cut into 1/2-inch dice
- salt and pepper to taste
- 1 bunch green onions, minced

Instructions:

1. Preheat the oven to 425°F (220 degrees C).
2. In a big saucepan of salted water, bring potatoes to a boil. Cook until the potatoes are cooked but still firm, approximately 15 minutes. Drain and transfer to a large mixing bowl.
3. Meanwhile, in a large saucepan over medium heat, cook the haddock. Pour the milk over the fish, followed by the green onions. Bring to a boil, covered. Reduce the heat to low and continue to cook until the salmon flakes easily with a fork.
4. Place the haddock on a platter. Remove and discard the bones and skin. Pour the milk mixture over the diced potatoes and stir well. Incorporate the cream fraiche. Season with salt and pepper to taste. Fold the haddock and half of the shredded Emmentaler cheese into the mixture gently. Fill a big, shallow baking dish halfway with the ingredients. Sprinkle with the remaining Emmentaler cheese on top.

5. Bake for 15 minutes in a preheated oven until the top is golden brown. Serve right away.

Dessert

Berry Crisp

Prep Time: 5 Minutes

Cook Time: 10 Minutes

Servings: 1

Ingredients:

- 1/4 cup fresh raspberries
- 2 tablespoons low-fat granola cereal
- 1/4 cup fresh blackberries
- 2 teaspoons light brown sugar
- Pinch of ground cinnamon

Instructions:

1. Preheat the oven to 400°F and coat an 8-ounce ramekin with cooking spray.
2. Stir together the berries, brown sugar, and cinnamon in the ramekin.
3. Sprinkle the granola cereal over the berry mixture.
4. 10 minutes in the oven
5. Serve hot.

Carrot Balls

Prep Time: 10 Minutes

Cook Time: 0 Minutes

Servings: 10

Ingredients:

- ¼ cup of raw walnuts
- 1 teaspoon nutmeg

- ¼ cup coconut, unsweetened finely shredded
- 1 carrot, finely grated
- 6 pieces Mejdool dates, pitted
- 1/8 teaspoon sea salt to taste

Instructions:

1. In a food processor, combine dates, 14 cup shredded carrots, salt, coconut, and nutmeg.
2. processor
3. Combine them and thoroughly process them.
4. Add walnuts and the remaining 14 cup carrots and blend until smooth.
5. a somewhat thick mixture
6. Form the mixture into balls and roll them in the shredded coconut.
7. Chill them in the fridge after topping with additional carrots.
8. Serve and have fun!

Cranberry Pistachio Biscotti

Prep Time: 25 Minutes

Cook Time: 45 Minutes

Servings: 36

Ingredients:

- ¼ cup light olive oil
- ¾ cup white sugar
- 2 large eggs
- 1 ½ cups pistachio nuts
- ½ cup dried cranberries

- ¼ teaspoon salt
- 2 teaspoons vanilla extract
- ½ teaspoon almond extract
- 1 ¾ cups all-purpose flour
- 1 teaspoon baking powder

Instructions:

1. Preheat the oven to 300° F. (150 degrees C). Use parchment paper to line a cookie sheet.
2. In a large mixing basin, combine sugar and oil until thoroughly combined. Mix in the vanilla and almond extracts, followed by the eggs. In a separate basin, combine the flour, baking powder, and salt; gradually fold into the egg mixture until mixed. Hand-fold in the pistachios and cranberries.
3. Cut the dough in half. On the prepared cookie sheet, shape each piece into a 12x2-inch log. Dough may be sticky; moisten your hands with cold water to make it easier to handle.
4. Bake until the logs are light brown, approximately 35 minutes in a preheated oven. Remove from the oven and leave aside for 10 minutes to cool.
5. Reduce the oven temperature to 275°F (135 degrees C).
6. Cut logs into 3/4-inch thick slices on the diagonal. Place the cut-side down biscuits on the cookie sheet.
7. Bake the cookies for 8 to 10 minutes, or until they are dry.

Coconut Bars

Prep Time: 10 Minutes

Cook Time: 30 Minutes

Servings: 4

Ingredients:

- ¼ cup cashews
- 1 teaspoon coconut oil
- 10 pieces dates, pitted
- ½ cup coconut, flaked
- 1/3 cup almonds, slivered

Instructions:

1. In a food processor, combine almonds.
2. Blend them until they are coarsely combined.
3. Add the dates and pulse.
4. Add the coconut oil and cashews and pulse until the mixture is thick.
5. Form the mixture into squares on wax paper.
6. Chill the squares for 20 minutes.
7. Serve and have fun!

Blueberry Muffins

Prep Time: 15 Minutes

Cook Time: 18 Minutes

Servings: 12 Muffins

Ingredients:

- 3 tbsp low fat margarine
- 1 tsp vanilla extract or vanilla paste
- ½ tsp baking soda
- 2 eggs medium
- 2 cups of fat free yogurt
- 2 cups blueberries
- 1.75 cups self-raising flour
- ⅔ cup superfine sugar

Instructions:

1. Preheat the oven to 375 degrees Fahrenheit / 190 degrees Celsius / 170 degrees Celsius (fan). Set aside muffin cases in a 12-hole muffin pan.
2. In a mixing dish, combine the flour and baking soda; keep aside until later.
3. In a separate dish, beat together the butter and sugar with a hand mixer until the mixture is light and fluffy. Continue to beat with the mixer after adding the eggs one at a time. If the mixture begins to split, add a couple of teaspoons of flour and continue mixing with a hand mixer until the mixture reaches a creamy consistency.
4. Combine the vanilla essence and fat-free yogurt in a mixing bowl.
5. Add the flour to the wet ingredients and gently mix in using a big spoon. Be cautious not to be too heavy-handed at this stage (although the mixture must be lump-free); just continue folding until the mixture has blended!
6. Fold in the blueberries gently.
7. Fill the muffin cases 34 full with the mixture using an ice cream scoop (or a spoon) and set on the middle shelf of the prepared oven.
8. Cook for 15 to 17 minutes, or until the muffins have begun to color and a toothpick inserted into the center comes out clean. Remove from the oven and set aside to cool on a wire rack before serving.

Pumpkin Muffins

Prep Time: 5 Minutes

Cook Time: 20 Minutes

Servings: 6

Ingredients:

- 1 can of pumpkin puree
- 1 box spice cake mix
- 1 cup of water

Instructions:

1. Preheat the oven to 350°F.
2. In a medium mixing basin, combine the spice cake mix, water, and pumpkin.
3. Fill each cupcake liner with the mixture and place in a cupcake pan. (If preferred, put a little brown sugar on top of each one.)
4. Bake for 20-22 minutes, or until a toothpick inserted into the center comes out clean.

Iced Pumpkin Cookies

Prep Time: 20 Minutes

Cook Time: 15 Minutes

Servings: 36

Ingredients:

- 1 cup canned pumpkin puree
- ½ teaspoon salt
- 2 ½ cups all-purpose flour
- 2 teaspoons ground cinnamon
- ½ teaspoon ground cloves
- 1 egg
- 1 teaspoon vanilla extract
- 1 teaspoon baking powder
- 1 teaspoon baking soda
- ½ teaspoon ground nutmeg
- ½ cup butter, softened
- 1 tablespoon melted butter
- 2 cups confectioners' sugar
- 3 tablespoons milk
- 1 ½ cups white sugar
- 1 teaspoon vanilla extract

Instructions:

1. Preheat the oven to 350° F. (175 degrees C). 2 cookie sheets, greased
2. Make some cookies: In a medium mixing bowl, combine the flour, cinnamon, baking powder, baking soda, nutmeg, cloves, and salt.

3. In a mixing bowl, cream together the sugar and butter until creamy, 2 to 3 minutes. Mix in the pumpkin, egg, and vanilla extract until smooth. Incorporate the flour mixture until well blended. Drop tablespoons of dough onto prepared cookie sheets; gently flatten.
4. Bake for 15 to 20 minutes, swapping racks halfway through, until centers are firm. Transfer the cookies to a wire rack to cool for approximately 30 minutes.
5. Meanwhile, prepare the icing: In a mixing dish, combine the confectioners' sugar, milk, butter, and vanilla extract until smooth. To create a dripping consistency, add milk as required.
6. Using a fork, drizzle frosting over cooled cookies.

Banana Shake Bowls

Prep Time: 5 Minutes

Cook Time: 0 Minutes

Servings: 4

Ingredients:

- ¾ cup almond milk
- ½ teaspoon vanilla extract
- 1 avocado, peeled, pitted, and mashed
- 4 medium bananas, peeled

Instructions:

1. Blend the bananas with the avocado and the remaining ingredients in a blender, pulse, divide into bowls, and chill.

Chocolaty Snickers Mini Tarts

Prep Time: 5 Minutes

Cook Time: 10 Minutes

Servings: 5

Ingredients:

- 10 filo mini shells, frozen
- Salt or chopped up peanuts for garnish
- One of snicker chocolate bar

Instructions:

1. Preheat the oven to 350 degrees Fahrenheit.
2. Lay out your filo shells on a baking pan.
3. Split the snicker bar in half.
4. Cut each half into five parts.
5. Fill the filo shells with them.
6. 10 minutes in the oven
7. Allow to cool after removing from the oven.
8. When finished, sprinkle with salt or chopped up peanuts.
9. Serve and have fun!

Blackberry and Apples Cobbler

Prep Time: 10 Minutes

Cook Time: 30 Minutes

Servings: 6

Ingredients:

- ¾ cup stevia
- ¼ teaspoon baking powder
- ¼ cup of apples, cored and cubed
- 1tbsp lime juice
- ½ cup almond flour
- ½ cup of water
- 3 and ½ tablespoon avocado oil
- 6 cups blackberries
- Cooking spray

Instructions:

1. In a mixing bowl, combine the berries with half of the stevia and lemon juice, then sprinkle with flour, stir, and pour into a baking dish sprayed with cooking spray.
2. In a separate basin, combine the remaining sugar, baking powder, water, and oil, and blend everything together with your hands.
3. Spread the mixture over the berries and bake for 30 minutes at 375 degrees F. Serve hot.

Conclusion

When you are trying to lose weight, it is hard to lose weight. You want to find a weight-loss plan that works for you and is easy enough to fit into your daily life. There are many different treatments, and while many may work well, others may be too hard or bad for your health.

This guide will look at Weight Watchers and how it might help your health. This diet plan is different from others because it doesn't have strict rules about what you can and can't eat. You can enjoy these pleasures sometimes, but you must be smart.

This book will explain how the Weight Watchers program works and how to use points and eat the right foods to live a healthy life. So, living a healthy life will help you lose weight, get more energy, and do many other things.

When you're tired of looking for the best diet plan that's easy to follow and helps you lose weight, it's time to look into the Weight Watchers program. This book will teach you everything you need to know to get healthy, lose weight, and have fun.

I hope you liked this book and got something out of it. I'd like it if you'd write a review of this book. We appreciate you buying this book.

The End

Made in the USA
Las Vegas, NV
08 December 2022

61572206R00072